FUZZY CONTROL OF INDUSTRIAL SYSTEMS

THEORY AND APPLICATIONS

FUZZY CONTROL OF INDUSTRIAL SYSTEMS

THEORY AND APPLICATIONS

by

IAN S. SHAW
Industrial Electronic Technology Research Group
Rand Afrikaans University, Johannesburg
Republic of South Africa

Kluwer Academic Publishers
Boston/Dordrecht/London

Distributors for North, Central and South America:
Kluwer Academic Publishers
101 Philip Drive
Assinippi Park
Norwell, Massachusetts 02061 USA

Distributors for all other countries:
Kluwer Academic Publishers
Distribution Centre
Post Office Box 322
3300 AH Dordrecht, THE NETHERLANDS

ISBN 978-1-4419-5055-0

Library of Congress Cataloging-in-Publication Data

A C.I.P. Catalogue record for this book is available
from the Library of Congress.

The front cover illustration has been derived from clip art published in Corel MegaGallery 2 and reproduced here with their permission.

NOTE TO INSTRUCTORS

A set of questions related to each chapter, suitable for classroom instruction or self testing, are available from the author free of charge by writing to: Author, P.O.Box 93519, Yeoville 2143, Republic of South Africa. Fax: +27-11-487-1880. e-mail: ishaw@global.co.za.

To Gladys...

CONTENTS

FIGURES

TABLES

PREFACE

This volume has been planned as an introductory textbook on intelligent control systems such as *fuzzy logic* and *neurofuzzy* systems. The objective was to create a linkage between an undergraduate text and a practical guide for experienced engineers wishing to upgrade their knowledge. To this end, both theoretical as well as practical design aspects are presented. However, basic familiarity with linear continuous and discrete systems, feedback, Proportional-Integral-Derivative (*PID*) controllers, and the principles of stability, dead time, frequency response, state space, would be necessary prerequisites. In contrast to linear systems, fuzzy and neurofuzzy control systems are rather specialized and the design engineer must be able to recognize situations when they would offer certain advantages over traditional methods, particularly in controlling highly nonlinear and time-variant plants and processes. The various off-the-shelf hardware and software-based fuzzy controllers described in trade journals and practically oriented seminars are relatively easy to understand and the mathematics involved are neither difficult nor obscure. However, both engineering students and practitioners need a deeper insight as to how fuzziness fits into modern control theory in order to gauge the accuracy, reliability and the limitations of their practical results.

Those interested in science in general and the role it plays in our everyday thinking will appreciate the philosophical and cultural implications of fuzzy logic. While not neglecting directly applicable knowledge, the classical goal of advanced technical education implies the engendering of a scientific world-view. In this process, some cultural preferences induced by axioms postulated hundreds of years ago might be superseded by something new that encompasses the old as a special case. It is not the first time that something like this happened. Think of relativity theory, the dual nature of light in physics, or the uncertainty relation in quantum mechanics as typical examples, each of which had forced a drastic change in our scientific world-view.

The special characteristics of fuzzy logic theory (also referred to as *possibility theory*) represent a novel way of handling uncertainty, quite different from *probability theory*. Possibility theory aims to predict the *degree* to which an event is happening, as opposed to probability theory which aims to predict the *chance* of an event happening. Furthermore, fuzzy theory can handle nonlinearities with ease, hence it provides a method to translate the vague, imprecise, qualitative, verbal expressions common in human communication into numeric values. This opens the door to convert human experience into a form understandable by computers. Thus fuzzy technology has an immense practical value because it allows to include the experience of human control operators into our current plant and process control design methodology. In addition, it provides a novel method to solve complex decision-making problems that often involve contradictory conditions. Neural networks and neurofuzzy systems represent a further extension by adding the capability of learning. Vague, uncertain, qualitative, verbal communication, learning and decision-making are distinctly human characteristics, hence fuzzy, neural and neurofuzzy techniques are often being referred to as being *intelligent* because they emulate human intelligence. The current worldwide success of fuzzy technology in industrial applications shows that it can contribute another useful design tool to the disciplines of industrial control engineering, manufacturing, human-machine communication and decision-making. It is time to get aboard!

The structure of this book is as follows.

In Chapter 1 the concepts of an intelligent system and the fundamental objectives of intelligent control are highlighted.

After reviewing the need for control system modeling and various modeling techniques in Chapter 2, more specific attention is devoted to fuzzy logic as an example of heuristic modeling. In turn, the basic principles of bivalence as an idealized world attribute and multivalence as a real-world attribute are discussed, along with the inherent imprecision of human communication which,, by necessity, attempts to express and manipulate the real-world. On this basis, the fuzzy implementation of intelligent control strategies are presented and some typical successful application examples are shown.

Chapters 3 and 4 present the basic theoretical framework of crisp and fuzzy set theory, relating these concepts to control engineering by pointing out the analogy between the *Laplace* transfer function of linear systems and the fuzzy relation of a nonlinear fuzzy system. As a result, the compositional rule of inference that uses the fuzzy relation to compute the output from a given input is shown as the fundamental rule that permits us to design nonlinear control systems based on fuzzy logic.

Chapter 5 deals with the generic structural aspects of fuzzy systems such as membership functions, fuzzification and defuzzification and gives practical guidelines as to how to handle the high degrees of freedom associated with fuzzy systems in choosing the most appropriate one from among the many techniques available.

Chapter 6 discusses three different fuzzy modeling techniques: rule-based, parametric and relational equation based. Although they are equivalent, each has certain advantages and disadvantages of which a fuzzy control design engineer must be aware. After a discussion of general problems related to industrial control, fuzzy *PID*, multivariable and supervisory control techniques and their practical realization are also discussed.

Chapter 7 discusses fuzzy systems identification, including adaptive and learning fuzzy systems. These techniques are particularly useful whenever there is no experienced human operator available and fuzzy controller parameters must be obtained from measurements

Chapter 8 takes a look at stability from a practical viewpoint and presents a method to determine practical stability limits for a fuzzy controller.

Chapter 9 presents an introduction to neural networks as a technology complementary to fuzzy logic which gives rise to neruofuzzy systems. It also gives a qualitative as well as a quantitative analysis of the backpropagation training algorithm which is the learning method used in neurofuzzy control systems. In particular, the trade-offs between obtaining fuzzy controller parameters from experienced human operators and/or measurements are examined in more detail.

Chapter 10 gives a detailed account of practical fuzzy controller design tools and discusses in detail the required features and practical uses of a generic fuzzy controller software design package.

Chapter 11 highlights several examples for successful practical industrial fuzzy controllers constructed and tested world wide. Its main objective is to act as an aid to the designer in recognizing potential applications for fuzzy control.

In as much as the original idea behind fuzzy control was the realization and inclusion of human control operator knowledge in automatic controllers, Chapter 12 examines classical and fuzzy human operator models to establish the optimum functioning of human operators, their capabilities and limitations in driving various industrial systems. An important part of this chapter is the validation of such models and the handling of intersubject variability.

Chapter 13 is an investigation of collaboration between various intelligent systems, including the information exchange between a human operator and a trained fuzzy controller. Some examples of implemented systems of the collaborative kind are described in more detail.

Chapter 14 draws general conclusions in the hope that future reasearch will greatly extend the practical application of fuzzy logic for industrial uses.

1 WHAT IS AN INTELLIGENT SYSTEM?

A so-called intelligent system gives appropriate *problem-solving* responses to problem inputs, even if such inputs are new and unexpected. Such behavior is "novel" or "creative". The workings of intelligent systems are usually described by analogies with biological systems by, for example, looking at how human beings perform control tasks, recognize patterns, or make decisions. At the moment, there exists a considerable mismatch between humans and machines, in as much as humans reason in uncertain, imprecise, fuzzy ways while machines and the computers that run them are based on binary reasoning. Eliminating this mismatch would make machines more intelligent, that is, they would be enabled to reason in a fuzzy manner like humans.

Notwithstanding its widespread use, the term "intelligent control" often causes confusion. *Intelligence* is often defined by the capacity to acquire and apply knowledge. On this basis, numerically controlled tool machines, microwave ovens, telephone switching networks and computers are "intelligent" in as much as they continually acquire knowledge (i.e. data) and apply them for various purposes. In fact, following any step-by-step algorithmic procedure can also be considered as intelligent at a certain level. Yet as regards human intelligence following a sequential logical procedure,such as an algorithm, is only one of its aspects which according to psychologists and neurologists, are governed by the left hemisphere of the human brain. Some refer to this form of intelligence as *objective intelligence* which yields quantifiable solutions to problems.

However, humans do a lot more. They also use intuition, past experience, guesses and hunches, rules-of-thumb, as well as recognizing and correlating over-all patterns instantaneously rather than sequentially. Such activities are said to belong to so-called right hemisphere brain functions. Experienced human control operators can translate process uncertainty into effective control action and can also explain their actions in an albeit ill-defined, imprecise, vague, qualitative linguistic manner. Furthermore, they can cope with *emergent* as opposed to *established* conditions which is the sign of creativity, i.e. of producing novel solutions to problems not encountered before. This kind of intelligence is also referred to as *subjective intelligence* which leads to non-quantifiable solutions to problems. It is clear that the subjective intelligence of *human operators makes a unique contribution which is not present in the objective quantifiable mathematical modeling intelligence that caters only to the logical mind.*

The question should arise: how can we include this human contribution expressed in vague, imprecise, qualitative terms in a mathematical model? A mathematical formalism was needed for the integration of qualitative and quantitative information, symbolic and numeric data, computation and human reasoning. This gave an impetus to the study of so-called artificial intelligence and its many disciplines like expert systems, fuzzy logic, neural networks, genetic algorithms and their various combinations.

Artificial intelligence is a discipline to study how humans solve problems and how machines can emulate intelligent problem-solving human behavior. In other words, how to make machines smarter by investing them with human intelligence.

Fuzzy logic is a technique to embody human-like thinking into a control system. A typical fuzzy controller can be designed to roughly emulate human *deductive* thinking, that is, the process people use to infer conclusions from what they know. For example, human operators can control complex nonlinear plants having poorly known dynamics. Fuzzy logic can capture their knowledge in a fuzzy controller to provide an equivalent performance to that of the human controller and, at the same time, adding a consistency and repeatability not present in a human controller.

In other fuzzy control approaches, the goal is to implement *inductive* rather than deductive systems. Such systems can learn and generalize from particular examples by observing current system behavior. This approach is referred to as *fuzzy learning control* or *fuzzy adaptive control.* Significant advantages may be obtained from controllers that can learn from experience, so that when a situation is encountered repeatedly, they will know how to handle the problem. On the one hand, *adaptive fuzzy systems* that can adjust to environmental changes have the ability to learn and explain their reasoning and, on the other, they have the capacity to be modified and extended. Such a balance between learned responsiveness and explicit human knowledge makes such systems very robust, extensible, and suitable for solving a variety of problems.

Controllers that combine intelligent and conventional techniques are also commonly used in the intelligent control of complex dynamic systems. *Embedded fuzzy controllers* where the fuzzy control function is only a portion of the total system's control strategy represent a typical example. Embedded fuzzy controllers automate what has traditionally been a human control activity. *Fuzzy supervisory industrial control systems* where the set points of many *PID* controllers are controlled by fuzzy means instead of human process operators represents another successful application area. In both of these cases human experience has been captured in a fuzzy controller. In summary, the fuzzy control method provides a heuristic technique for constructing nonlinear controllers.

Another intelligent control area is the application of *artificial neural networks* which emulate low-level biological functions of the brain to solve difficult control problems. Neural networks have the capability to learn from experience how to model and control a system.

The ultimate goal of intelligent system design is the creation of *autonomous systems* which can perform complex control tasks under all operating conditions of a plant or process, even in the presence of failures, without human intervention or supervision. In this case it will be sufficient to tell the system *what* to do but *not how* to perform the task given.

References

APOSTEL, L: "Formal Study of Models." In *The Concept and the Role of the Model in Mathematics and Natural and Social Science.* Freudenthal H, ed. Kluwer Academic Press, Dordrecht, 1961.

ARNOLD WR, BOWIE JS: *Artificial Intelligence - A Personal Commonsense Journey.*Prentice Hall, Englewood Cliffs, NJ , USA, 1986.

BARR ,A., FEIGENBAUM, EA: *The Handbook of Artificial Intelligence.* Los Altos, Ca. 1981; 1; 1982; 2-3.

FORDYCE, K., NORDEN, P. , SULLIVAN, G: "Artificial Intelligence and the Management Science Practitioner: One Definition of Knowledge-Based Expert Systems." Interfaces, 1989;19; 66-70.

HARMON P., KING D.: *Expert Systems: AI in business.* John Wiley and Sons, New York, 1985.

KASTNER, JK., HONG, SJ: "A Review of Expert Systems."European Journal of Operations Research. 1984; 18; 285-292.

2 MODELING PLANTS AND PROCESSES FOR CONTROL SYSTEMS

2.1 The main tasks of an industrial control system

It is useful to remind ourselves of the main tasks of an industrial control system which is liable to operate in a noisy and polluted environment subject to unpredictable disturbances such as fluctuating temperatures, humidity, and electrical line voltage transients. The first task of an industrial controller is to suppress the influence of such external disturbances by changing the over-all system characteristics in order to compensate for the unfavorable effects mentioned. Secondly, the controller must ensure the stability of the plant or process under varying operational conditions. If as a result of some an external factor one of the state variables deviates from its operating point but returns to it in time, then the process is deemed stable. The controller must influence the process such that this will happen under all operating conditions. Thirdly, the controller must ensure the optimum performance of the plant or process by producing the largest amount of the output product. Maximum productivity is particularly important in the chemical industry where profits depend on the quantities of output produced. It is clear that the control system designer must know the operating characteristics of the process as well as all possible operating conditions in order to carry out his work successfully. This process knowledge can take various forms, as outlined in the following sections.

2.2 The need for modeling

Models of physical reality are desirable because they can be manipulated much easier and cheaper than physical reality itself. In the following, some modeling methods of control systems will be reviewed.

2.3 The experimental method

Consider an existing single-input single-output memoryless system whose behavior is characterized by an input-output table (Table 2.1) constructed experimentally by measuring the output responses to a set of input values. (The situation becomes somewhat more complex if there are time constants and/or time delays between inputs and outputs. In this case the contents of the output column will be a function of time). Graphically the method is equivalent to plotting some discrete points of the input-output curve, using the horizontal axis for input and the vertical axis for output. Often the physical equipment of the process has not yet been constructed. Consequently one cannot experiment to determine how the process reacts to various inputs and therefore one cannot design the appropriate control system. But even if the process equipment were available for experimentation, the procedure would usually be very costly. Besides, for a large number of input values it would be impractical to measure the output and interpolation between measured outputs would be required. One must also be careful to determine the expected ranges of inputs and outputs to make sure that they fall within the range of the measuring instruments available.

2.4 The mathematical modeling method

It would be more desirable to be able to calculate, that is, predict all possible output values from all possible input values on paper, without resorting to measurements. Therefore one would need a simple description of how the process reacts to various inputs and this is what mathematical models can provide for the designer. The conventional approach to control engineering requires an idealized mathematical model of the controlled process, usually in the form of differential or difference equations (or *Laplace* and *z*-transforms respectively). In order to make mathematical models simple enough, certain assumptions are made, one of which is that the process is *linear*, that is, its output is proportional to the input. Linear techniques are well-developed and their main value is that they provide good insight. Besides, there exists no general theory for the analytic solution of nonlinear differential equations and consequently no comprehensive analysis tools for nonlinear dynamic systems.

Table 2.1. Input-output relationships

INPUTS	OUTPUTS
0	1.5
2.6	4.5
4.7	7.9
9.1	15.3
12.6	29.0

Explicit analytic solutions are usually available only for linear differential or difference equations (in cases, whenever the operating point changes only within relatively narrow limits, a nonlinear system can be adequately approximated by a linear system). Another assumption is that the process parameters do not change in time (that is, the system is *time-invariant*) despite system component aging, wear and tear, and environmental changes. Due to such simplifications, one might encounter serious difficulties in developing a meaningful and realistic mathematical description of an indutsrial process. The causes of such difficulties may be classified as follows:

Poorly understood chemical or physical phenomena

To understand completely the physical and chemical phenomena occurring in a process is virtually impossible. Even an acceptable degree of knowledge is at times very difficult to achieve.

Inaccurate values of various parameters

Accurate values of model parameters are indispensable for any quantitative analysis of process behavior. Unfortunately, they are rarely available. In addition, parameter values change with time and one needs some quantitative description as to how this occurs. The mathematical approach finds it difficult to cope with the *uncertainty* of parameter values and their changes due to environmental variations. The *dead time* is also a critical parameter whose value is either imprecisely know or varies with time. Poor knowledge of dead time and its variation can lead to serious stability problems.

The size and complexity of the model

In an effort to develop an accurate and precise mathematical model, its size and complexity tends to increase significantly. Care must be exercised that the size and complexity of a model do not exceed certain manageable levels, beyond which the model loses its value. In accordance with this, nearly *80%* of all industrial controllers used today are of the *PID* type and complex multivariable controllers based on advanced control theory are much under-represented. However, *PID* controllers are linear controllers and they are not suitable to drive strongly nonlinear plants or processes. Recent surveys by Brown [1994] in South Africa and the United States have indicated that an inordinate number of existing *PID* controllers are not properly tuned and in many installations they were even found to be used in the *manual mode.* The question is whether or not this is due to the necessity for frequent re-tuning of *PID* controllers used for controlling nonlinear plants for which, by their very nature, are not suitable.

External disturbances

External disturbances expected to appear will influence the mathematical model. Those with a small impact on process operation can be neglected, whereas others with significant influence must be included in the model. The decision as to which variables should be included will impact upon model complexity.

Shortage of skilled manpower

Although throughout the years major contributions have been made to advanced control theory which go much beyond the *PID* controllers in use in most plants today, these did not seem to have made much impact on industrial control. This is most likely attributable to the lack of sufficiently skilled manpower. In fact, it seems that *PID* controllers represent the limits of complexity understood in most industrial plants, while equipment utilizing advanced control theory would require a higher level of mathematical understanding. Personnel with such capabilities is usually not a part of the average maintenance team. It is regrettable that this very relevant problem has never been seriously addressed by the leading world authorities in control systems[1]. Yet the unique property of intelligent controllers, and especially of fuzzy controllers, is that they are based on human experience and thinking patterns rather than mathematical models. This means that the training of control operators and maintenance technicians is expected to be easier and a lot less expensive, and that less qualified personnel needs to be used. *This factor alone should become a very important argument in deciding whether or not to adopt fuzzy controllers in specific applications.*

As expected, simplifying assumptions in mathematical models may result in significant information being ignored and this loss must be rectified later on by the *tuning* and adjustment of the controller in a real plant or process operation. If done in a sufficiently careful and thorough way, this method certainly works in many applications. However, when the process complexity approaches or exceeds a certain threshold, mathematical models not only become far too complex and intractable, but their accuracy and reliability in approaching the physical reality they attempt to model also become highly questionable. In fact, advanced control theory itself has gained such a mathematical complexity, that it has been completely divorced from physical reality. As a result, many people think today that control engineering is a mathematical rather than an engineering problem. Practicing engineers have long since come to the conclusion that because of the many simplifications made in mathematical modeling, it is fallacious to strive for more and more modeling accuracy. Yet the notion of mathematical accuracy has been so deeply ingrained, that until the mid-*60*'s no self-respecting scientist or engineer dared to challenge it. This is, of course, not surprising. It is said: "The aims of scientific inquiry and technological implementation are not the same. Science delights in complexity as it discovers ever more subtle phenomena. Technology abhors complexity as it tends to render the product in question more difficult to use and maintain." The fact that mathematics, which is a self-consistent product of human thinking, can be used so effectively to produce representations of physical reality is a remarkable phenomenon itself. However, more recent thinkers consider mathematical models as a bridge between thinking and experimenting rather than the only true measure of describing the physical world.

2.5 The heuristic method

The heuristic method consists of performing a task in accordance with previous experience, tips, hunches, rules-of-thumb and often-used strategies. A heuristic rule is a logical implication of the form:

If <*condition*> then <*consequence*>

or in a typical control situation:

If <*condition*> then <*action*>

Rules associate conclusions (or consequents) with conditions (or antecedents). The heuristic method is actually similar to the experimental method of constructing a table of inputs and corresponding output values. For each row of Table 2.1 we have a relationship between input and output variables. We can write a heuristic rule for a row, such as, for example:

If $e_{in} = 1$ volt then $e_{out} = 3.5$ volts

Instead of crisp (i.e. non-fuzzy) numeric values of input and output variables, one can also use linguistic (i.e.verbally expressed) values:

If $e_{in} = MEDIUM$ then $e_{out} = LARGE$

where *MEDIUM* and *LARGE* are defined linguistic functions. Note that instead of a numeric expression, *linguistic* (i.e. verbal) expressions were used. These linguistic expressions are also called *fuzzy* expressions because they represent classes of uncertainties. More importantly, the assumption of linearity is no longer necessary or even relevant, since the plant or process input-output function is described point-by-point, just like in the experimental method. In other words, the power of the heuristic method lies in its ability to construct a useful *non-mathematical* input-output function for a plant or process whenever an equivalent mathematical model would either be too complex, or would have uncertain or even unknown parameters., According to Kosko [1992] in the linguistic fuzzy approach a real-world input-output function is approximated by overlapping patches and the position of the point representing the true value of the function is uncertain within a patch. Instead of specifying an exact value of a point of the input-output function, a patch specifies a whole set. The bigger a patch the more uncertain (i.e. the more fuzzy) it is as to which value in such a set represents the real point. Each patch represents a heuristic fuzzy **IF…THEN** rule. Conversely, the less fuzzy the rules, the smaller the patches. If the rules contain crisp numbers rather than fuzzy sets, the patches collapse into points. According to the *Fuzzy Approximation Theorem* (see Chapter 6) a curve can always be covered with a finite number of patches. It will be seen later that the above rule structure is used explicitly by some so-called *intelligent systems* such as expert systems and fuzzy systems, including neurofuzzy systems (see Chapter 9).

Fuzzy logic techniques translate qualitative linguistic statements about control procedures into computer algorithms. A fuzzy logic controller is a nonlinear algorithm representing the qualitative knowledge of a human expert about system behavior and the desired control action. Thus fuzzy systems are:

- knowledge-based systems constructed from expert knowledge
- universal approximators that can realize nonlinear mappings

This dual nature allows qualitative knowledge to be combined with quantitative data in a complementary way.

2.6 Why is fuzzy logic needed?

New technologies are invented out of specific needs. The advent of fuzzy logic was precipitated by the need for a rigorous method, capable of *expressing imprecise, vague, and ill-defined quantities in a systematic manner*. For example, instead of a mathematical model, fuzzy logic based industrial controllers utilizing heuristic rules can be infused with the experiential knowledge of a trained human operator, yet the control action can be as good, often better, and always more consistent. Trained human control operators can cope with ill-defined or poorly understood plants or processes even if they do not understand their underlying dynamics. They know what *action* to take whenever they observe certain *conditions,* such as combinations of instrument readings, indicator lamp patterns, or other signals, even if they do not have a deep understanding of the plant or process. This solves the problem of automating control functions hitherto only possible by manual means. Fuzzy logic is also useful in complex *decision-making tasks* where the individual variables are not definable in exact terms. Such cases are, for example, industrial production scheduling as well as logistics and maintenance planning where the use of fuzzy logic can realize significant cost advantages.

Fuzzy logic has not been easily accepted by everyone, in as much as it goes against our Western cultural and scientific traditions firmly rooted in mathematical precision, bivalent logic and the abhorrence of human subjectivity . Although these traditions have served us amazingly well considering the difficulties of problems faced in the past, it has become necessary to fill the gaps not adequately addressed by traditional methods. It will be seen in later chapters, that conventional logic is but a special case of fuzzy logic, hence it is to be expected that the judicious combination of conventional and fuzzy control methods will ultimately yield the best practical results. As has been mentioned, fuzzy logic, neural networks, and expert systems belong to the new paradigm of *intelligent systems.* As expected, any new technology has its proponents and opponents. In deciding whether or not to adopt fuzzy technology, seasoned engineers should be guided by the requirement of evaluating it on the basis of usefulness, reliability and economic viability. Thus it is essential to ask the following questions:

- **Is fuzzy logic only a "buzzword" technology or does it offer objective and tangible advantages?**
- **Does fuzzy logic enable me to solve problems which were hitherto either too difficult or even impossible to solve?**
- **Is fuzzy logic economical to use?**
- **What level of training is required to maintain and repair fuzzy controllers?**
- **Do I understand in which specific applications the advantages of fuzzy logic could be realized?**

The aim of this work is to provide answers to these questions[2].

2.7 Conventional versus intelligent control modeling

In conventional control methodology shown in Figure 2.1, *what is modeled is the plant or process being controlled.* In this procedure called *systems identification*, the system is assumed to be linear or nearly so, characterized by a set of differential equations whose solution tells the controller how its parameters should be adjusted for each type of systems behavior (such as damping, overshoot, speed of response, settling time and steady-state error) required. On the other hand, in many systems not amenable to automatic control, human control operators are still being employed and intelligent control methodology focuses on such a human operator's behavior, that is, how he/she would adjust the control parameters for a given set of circumstances. In fuzzy methodology *it is the operator whose model is being identified* while he/she is controlling the system, as shown in Figure 2.2. Thus the fuzzy controller, based on the thus identified human operator model, becomes a logical model of the thinking process a human operator might go through when manipulating the system. *This shift in focus from system to person changes the entire approach to automatic control problems.*

2.8 The right controller for the right application

Intelligent control is not a panacea and one would certainly be wrong in advocating the mindless replacement of conventional controllers with fuzzy controllers. It is therefore useful to look at the relative applicability of both conventional and fuzzy controllers. If the plant or process being controlled is not strictly linear but the nonlinear input-output function is smooth and contains no abrupt jumps, or whenever the operating point is fairly stable with environmental changes, then within a small range around the operating point any change in the input variable causes an approximately proportional change in the output variable. In such cases the *PID* controller (or rather a programmable logic controller (*PLC*) programmed to function as a *PID* controller) is still the most cost-effective choice. As long as these conditions prevail, *PID* controllers can control plants even with incompletely known dynamics, in as much as the *P*-component represents the feedback error at a particular instant, the *I*-component being the integral of the error contributes the past history of the feedback error, while the *D*-component being the derivative of the error attempts to anticipate the future behavior of the feedback error. If the parameters of each component are tuned for the specific performance needs of the plant, the controller action will be satisfactory. *Tuning* implies the mutual optimization of response characteristics such as damping, overshoot, settling time, and steady-state error (or offset). Of course, only a compromise between these contradictory control goals can be achieved. Linearity (or quasi-linearity) guarantees that then three individual control strategies can be combined in an additive manner, whereas non-linearity would cause an interaction between them, thereby making tuning difficult or even impossible. Otherwise, the feedback loop itself should adequately compensate for plant parameter changes, noise and environmental are used to dynamically adjust the time behavior of the feedback changes. Thus the *PID* controller represents three specific control strategies which error function. In as much as *PID* controllers are single-input single-output controllers and most industrial plants and processes are of a multivariable nature, one can readily see that each controlled variable requires its own control loop and set point. Hence a need arises for a ***supervisory control system*** capable of adjusting the set points of a multitude of control loops (Refer to Chapter 6). In many systems this can be accomplished by the time-scheduling of set point adjustments, and computers can be used. In others, the change of set points must be done on the basis of decisions dictated by the observation of current process outputs, and human judgment must still be employed. If the plant or process dynamics are substantially nonlinear and the operating point shifts over a considerable range, or the plant parameters change considerably due to environmental variations, then *PID* controllers cannot provide satisfactory performance. Unfortunately the range of industrial control variables is usually very wide and nonlinear, while plant dynamics are prone to parameter changes and are imprecisely defined. In many processes, such as, for example, rubber extrusions and automobile tire manufacturing, the input-output function of a ***Banbury mixer*** or side-wall extruder cannot be described mathematically becausethe related complex reactions and process dynamics are not fully understood even by experienced operating personnel. There exists, however, a large body of ***empirical recipes*** which are known to produce acceptable results.

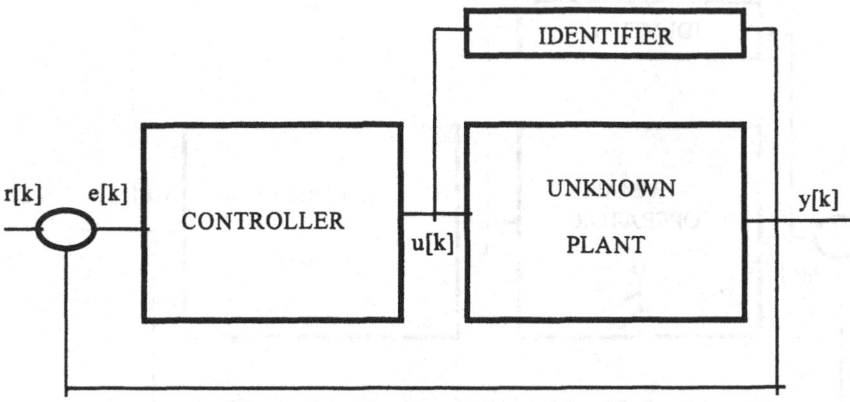

Figure 2.1. The plant is being identified

These recipes can also be regarded as input-output functions because they relate key input variables to output variables even if they have incomparable dimensions such as, for example, input pressure and material composition versus tire lifetime. Such systems are amenable to fuzzy control. One of the important feature of fuzzy controllers is their ability to carry out multi-objective control whereby several different, even conflicting requirements are reconciled to achieve a compromise control strategy.

A good example is a container crane where a load suspended from a chain must be transferred from a ship to a railroad car. In the simplified version shown in Chapter 11, the two input variables of the controller are: the *distance* between the ship and the railroad car, and the *angle* of the chain by which the load will sway, while the output variable is the *motor power* that drags the load from its initial point to its destination. A human operator would observe both the angle and distance variables simultaneously and adjust the speed of transfer (by means of the motor power controller) such as to effect the fastest possible transfer speed while keeping the swaying angle within reasonable limits. Theoretically speaking, by using appropriate sensors (see comments in Chapter 11 on practical sensors in this respect!), a two-input single-output fuzzy controller would be capable to emulate this multi-objective control action of the human operator. It is clear that it would be difficult to use two single-input single-output *PID* controllers for this purpose because of the lack of suitable coordination between controllers assigned to each variable. Conventional *PID* and fuzzy controllers are sometimes combined, for example, in some commercial temperature controllers. In order to provide fast control action, the *PID* controller is tuned for fast rise time that would normally cause a substantial overshoot. Just before the inception of the overshoot, the fuzzy controller cuts in and flattens the response at the set point. This is a particularly judicious combination of conventional and nonlinear techniques using fuzzy control and has been incorporated. in several products. A considerable number of fuzzy controllers are being used as so-called *embedded* controllers whereby they take over only some of the functions of a conventional controller.

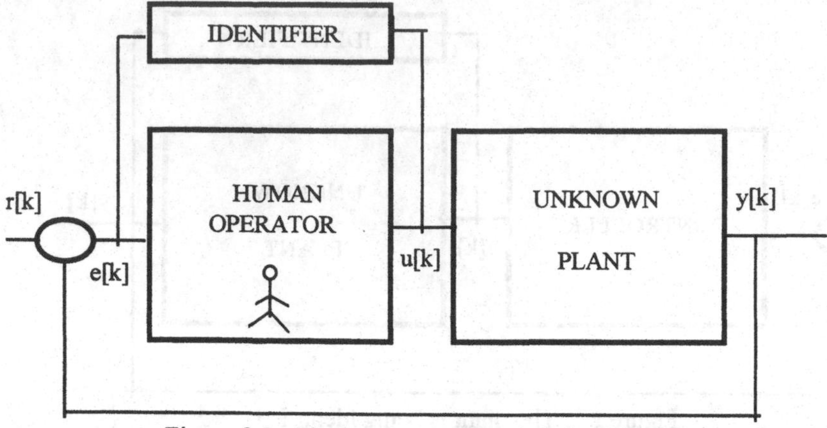

Figure 2.2. The human operator is being identified

.In fact in has been estimated that by the turn of the century over *60* % of all fuzzy controllers will be embedded controllers. One of the reasons is that embedded controllers can be packaged as self-contained modules containing microprocessors and read-only memories loaded with software that executes fuzzy control functions, which could be used in many applications.

2.9 Fuzzy implementations of intelligent control strategies

Although intelligent control strategies may be implemented by means of other than fuzzy logic, fuzzy implementations often turn out to be more efficient, for the following reasons:

- Fuzzy control strategies stem from experience and experiments rather than from mathematical models. Hence a linguistic implementation is much faster.
- Fuzzy control strategies involve a large number of inputs most of which are relevant only for some special conditions. Such inputs are activated only whenever that specific condition prevails. In this way, some rare or exceptional (such as alarm) conditions can be incorporated with very little additional computational overhead, yet the rule structure of even very complex systems remains transparent and understandable.
- Fuzzy logic strategies implemented in mass-market products must be implemented cost-effectively. Compared with conventional solutions, fuzzy logic is often more efficient as regards coding efficiency and computational overhead.

2.10 Typical successful applications of fuzzy control

This subject will be discussed again in Chapter 11. However, a few salient examples are worth mentioning here.

- Whenever machine setting with the aim to reduce waste and scrap are a matter of *operator judgment*. An example is saw milling due to the variability of logs. Fuzzy control provides a means for embedding experiential knowledge.

- Whenever the system depends on *operator skill and attention*. An example is coke production where even the most experienced operator cannot maintain a consistent product quality.
- Whenever one process parameter affects another process parameter. One example is the container crane already mentioned where the operator must simultaneously observe transfer speed and the angle of swing. Another example is injection molding where nozzle temperature and screw pressure are interrelated: resin flow is dependent on both temperature and pressure. In such applications fuzzy control creates a bridge of interaction between the variables and achieves the best compromise. This is an effect that cannot be attained by separate *PID* control loops, yet a human operator (or a fuzzy controller implementing his empirical knowledge) can cope quite well.
- Whenever processes can be modeled *linguistically* (i.e. by qualitative verbal description) but not mathematically. An example is tension control of film or paper by means of a tension roller ("dancer") whose position is described linguistically using operator judgment.
- Whenever a fuzzy controller can be used as an advisor to the human operator. Given the empirical nature of fuzzy controller design, there is often a desire for extra caution, especially when a new fuzzy controller is being introduced in a practical installation currently employing a human control operator. It is possible to display on a screen the fuzzy controller output only as a guidance to the human operator who executes the actual control functions. The operator usually finds the fuzzy controller output as being reasonable, based on his experiential knowledge of the process. However, he/she might not have used the same process control inputs consistently in the same manner.

2.11. Model validation

Any model must be properly validated as, for example, by the method shown in Figure 2.3. The essence of the method is that the same input signal $\boldsymbol{u}[k]$, a discrete time vector, is applied to both the real system to be modeled and the model itself. The comparator shown generates the modeling error $\boldsymbol{e}[k]$ which is the absolute value of the difference between the real system's output $y[k]$ and its estimate $y\hat{}[k]$.[3] In general, given k discrete observations of random signal vector $\boldsymbol{u}[k]$ we wish to design a signal processor that gives us the optimum estimate $y\hat{}[k]$ of the related signal vector $y[k]$ in a finite time interval $k_a \leq k \leq k_b$. Thus we are seeking the function $y[k] = f(u[k])$ for each discrete k. Assume that the optimum signal processor utilizes the linear mean square (*LMS*) criterion, i.e. the estimate is to be a *linear* function of the observations:

$$y\hat{}[k] = \sum_{i=k_a}^{k_b} h(k,i)\, \boldsymbol{u}_i \tag{2.1}$$

For each k, the weights $h(k,i)$, $k_a \leq i \leq k_b$ are selected so as to minimize the mean-square estimation error:

$$E = E\,(e^2\,[k]\,) = E\,(y[k] - y\hat{}[k]\,)^2 = \text{minimum} \tag{2.2}$$

where E denotes expected value. The signal processor which determines the optimum weights $h(k,i)$ according to the mean square criterion is in general referred to as the *Wiener estimator*.

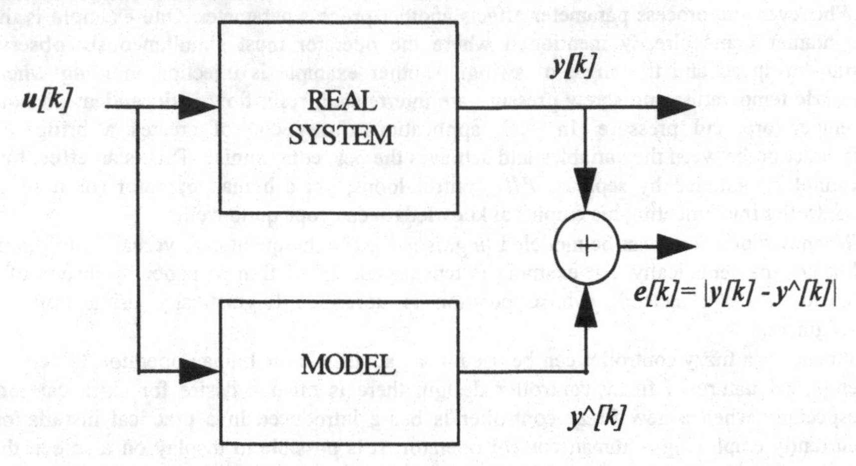

Figure 2.3. Model validation.

Batched estimators

Batched estimators usually solve three types of problems: *smoothing*, *filtering* and *predicting*. This division into three types of problems depends on the specific portion of the interval $[k_a, k_b]$ from which the observed samples u_i sre taken to determine $y\hat{}_i$

Assume that the data is arriving in a batch form, $\{ u[k_a],..., u[k_i],...,u[k_b] \}$. If the model output $y\hat{}[k]$ also exists for the same batch length $k_a,...,k_b$, then the estimate is called *smoothing* because the same data batch used for generating the real system output, $y[k]$, is being re-estimated by the model and compared for model errors. Figure 2.4 shows the smoothing problem where all observations in $[k_a, k_b]$ are taken into account, i.e. a whole batch is processed. Since some observations are what amounts to be the future of $y[k]$, the operation is not causal, the data batch $u[k]$ must be stored in memory prior to commencing the operation.

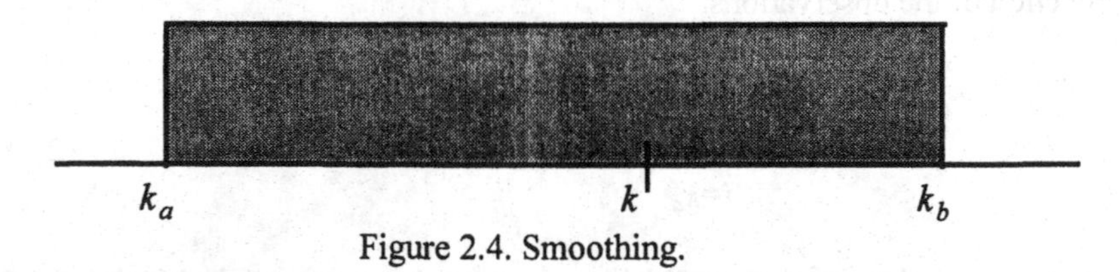

Figure 2.4. Smoothing.

Figure 2.5 shows the *filtering* problem where only the current sample and all past samples are used in making up the estimate $y\hat{}[k]$. Thus only the shaded portion of the observation interval is used at the current time instant and the estimate is calculated as:

$$y\hat{}[k] = \sum_{i=k_a}^{k} h(k,i)\, u_i \tag{2.3}$$

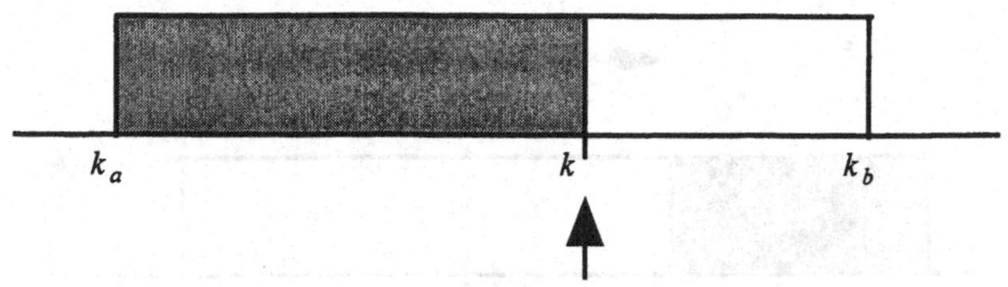

Figure 2.5. Filtering.

Note that the estimator acquires data during the interval $[k_a , k_b]$ and makes its estimate on this basis for $[k+1, k_b]$. Another version of this estimator is a *moving average filter* which takes into account only the current observation and the past M observations: u_i ; $i = k, k-1,...,k-M$. That is, only $M + 1$ filter weights must be computed at each sampling instant k. This procedure is shown in Figure 2.6 and the estimate is calculated as:

$$y\hat{}[k] = \sum_{i=k-M}^{k} h(k,i)\, u_i = \sum_{m=0}^{M} h(k,k-m)\, u_{n-m} \tag{2.4}$$

This type of estimator is referred to as the *finite impulse response* (*FIR*) filter. Note that the estimator acquires data in the interval $[k-M, k]$ and as the batch M moves forward with each new sample, the oldest sample is "forgotten".

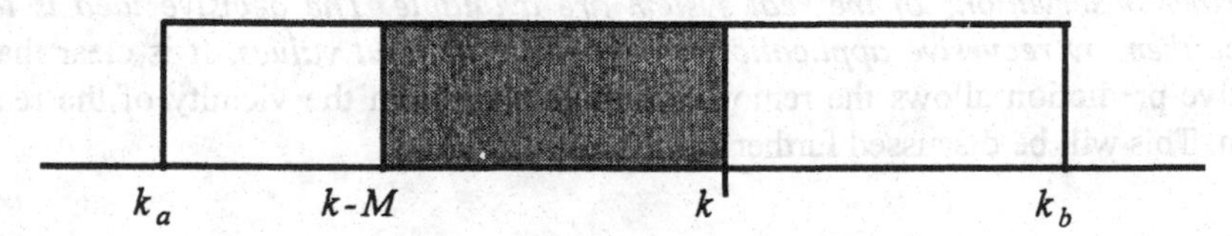

Figure 2.6. Moving average filter.

Figure 2.7 shows the problem of *predicting* which is a special case of filtering in that observations only up to the instant $k - D$ must be used in obtaining the current estimate $y\hat{}[k]$. This is equivalent to predicting the futuire D samples. The estimate is calculated as:

$$y\hat{}[k] = \sum_{i=k_a}^{k-D} h(k,i)\, u_i \tag{2.5}$$

Note that the estimator acquires data during the interval $[k_a, k\text{-}D]$ on the basis of which it makes future predictions. This type of estimator also has another version, the *moving average predictor*, where the prediction is based only on the last M samples. This version will not be discussed here.

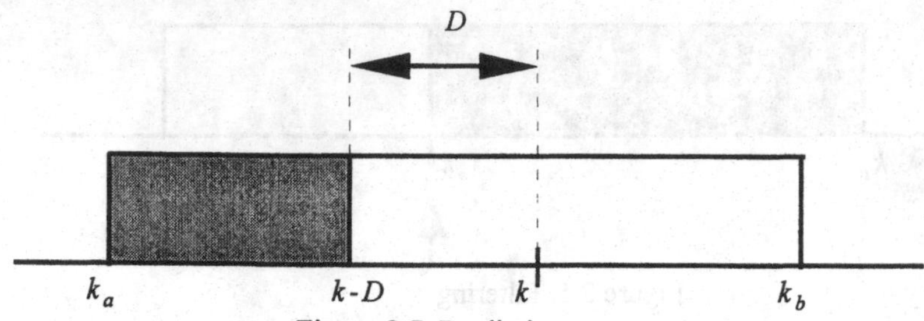

Figure 2.7. Prediction.

Recursive operation

In recursive operation, instead of a batch of data, at most only a few data items are stored. Assume that an initial estimate $y\hat{}[k_{init}]$ can be generated by some means and then stored. The new incoming data sample which becomes available at the following sampling time is used together with the initial estimate (which is now $y\hat{}[k_{init}-1]$) to generate a new estimate as follows: (assume for the moment that K is known):

$$y\hat{}[k] = y[k_{init}] + K\,(y[k] - y\hat{}[k]) \tag{2.6}$$

In this particular case, the new estimate can be stored in a single-cell memory and the process is continued in the same fashion. We speak of prediction whenever the time of interest occurs after the last available observation, hence our estimator operates from here on as a predictor. Note that by definition in the prediction stage *no further observations of the real system are available. The decisive idea is to replace them by recursive applications of predicted output values.* It is clear that recursive prediction allows the removal of the model from the vicinity of the real system. This will be discussed further in Chapters 6 and 12.

Notes

1. In one of my lectures to a university staff, I mentioned that the average factory technician cannot cope with the maintenance, repair and calibration of advanced controllers based on complex mathematics, while fuzzy controllers would alleviate this problem because their structure is analogous to human thought patterns. One of my critics in the audience said that this is a rather poor justification and all I am doing is trying to fix a neglect by introducing a new design paradigm. According to him, the logical approach would be to educate technicians so that they can understand the complex mathematics of advanced controllers. Although there is some truth to this argument, one should always remember that control engineering is a human activity and as such, it encompasses educational, social and psychological issues that go far beyond the technical contents of industrial employment.

2. In this work we shall deliberately avoid the debate between those that advocate probability theory as the only true analysis tool for uncertainty, and those who argue that fuzzy theory is the more natural representation. Our aim is to examine the connection between the theory and the practical application of fuzzy and neurofuzzy systems in industrial control. The worldwide success of hundreds of practically

implemented fuzzy systems constitutes an ample proof of the fuzzy paradigm and its usefulness as a viable tool of control engineering.
3. In industrial systems, other performance criteria such as various time-integrals are used. In terms of continuous time functions, the error is $e(t) = y_{setpoint}(t) - y(t)$ and the most often used time integrals with integration limits *0* to ∞ are: integral of the square error: $ISE = \int e^2 (t) dt$; integral of the absolute value of the error: $IAE = \int | e(t) | dt$; Integral of the time-weighted absolute error: $ITAE = \int t |e(t)|$. If large errors are to be suppressed, ISE is recommended because the errors are squared and thus contribute more to the value of the integral. If the errors to be suppressed are small , IAE is better than ISE because when we square small numbers (smaller than one), they become even smaller. To suppress errors that persist for long times, ITAE will tune the controllers better because the presence of large t values amplifies the effect of even small errors.

References

BROWN M: "The State Of *PID* Control In South Africa". SA Instrumentation and Control, August, 1994.
CASTRO JL:"Fuzzy Logic Controllers Are Universal Approximators.IEEE Trans.Sys.,Man,Cybern.,1995;25;4;629-635.
KOSKO B: *Neural Networks and Fuzzy Systems*. Prentice Hall, 1992.
MAMDANI EH: "Twenty Years of Fuzzy Control: Experiences Gained and Lessons Learned." 2nd IEEE Internat'l Conf. on Fuzzy Systems, ISBN 0-7803-0615-5, 339-344.
PASSINO KM:"Bridging the Gap Between Conventional and Intelligent Control." IEEE Control Systems Magazine, 1993;8;12-18.
"The Fuzzy Logic Market 1991-1996". SA Instrumentation and Control, Electronic Trends Publications, September 1993;21.
STEPHANOPOULOS G:*Chemical Process Control.*Prentice Hall, 1984
TONG RM: "A Retrospective View of Fuzzy Control Systems." Fuzzy Set and Systems, 1984;14.
ZADEH LA: "Outline Of a New Approach To the Analysis of Complex Systems and Decision Processes." IEEE Trans. on Sys,Man, Cybern.,1973; 3; 28-44

3 INTRODUCTION TO SET THEORY AND FUZZY LOGIC

3.1 The set-theoretical approach to systems

Given a multi-input multi-output system with *m* input and *p* output terminals. Applied to these terminals is a set of possible input and output values respectively. In order to obtain the combined input to the system, one must be able to know how to combine the *m* input sets. A similar reasoning applies to the *p* outputs where at every output terminal a different set of output values can arise. To determine the total output, one must be able to know how to combine the *p* output sets. For this reason, one must study the appropriate *set-theoretical operations* whereby to combine two or more sets.In systems analysis and design one is also interested in how the input affects the output; in other words, how the input is mapped into the output by the given system. In other words, how the system transforms the combined *m* input sets into the *p* output sets. The *set-theoretical operations* that provide this information establish an input-output mapping analogous to the transfer function of linear system theory.

3.2 Basic notions of sets

The basic idea in set theory is the so-called *membership of an element* x in a *set* A using the symbol $\in$:

$$x \in A \tag{3.1}$$

In order to indicate this membership, one can use a *membership function* $\mu_A(x)$ whose value indicates whether or not the element x is a member of the set A.
For example, in the following case $\mu_A(x)$ is a bivalent function

$$\begin{aligned} \mu_A(x) &= 1 \ \ if\, x \in A \\ &= 0 \ \ if\, x \notin A \end{aligned} \tag{3.2}$$

Figure 3.1 illustrates these cases. One observes, that if element x_2 is moved towards the boundary of set A, at the point of crossing it there will be a sudden jump in its membership value from non-member to member. Another observation is that the membership of an element which is right on the boundary is undetermined. As an example, consider a *TTL* binary logic gate with its output defined nominally at *0* Volts for logic "*0*" and *5* Volts for logic "*1*". If the gate is faulty, it might "hang" somewhere between these values in an undefined state.

The fundamental property of conventional (so-called *crisp*) logic is that a membership function is bivalent.

In conventional set theory, a set of elements, for example, are values indicated on an instrument scale: "*0....100* Volts". A given value may be an element of this set (such as, for example "*25* Volts") , or it may not be an element of the set (such as, for example, *134* Volts"). Let us, however, take another example from everyday life. Consider Figure 3.2 where speed is plotted against the bivalent membership function of Equation 3.2 and the speed limit is shown as *70* km/h. Those who drive faster than *70* km/h belong to the set A whose elements are violators and their membership function has the value of *1*. On the other hand, those who drive slower do not belong to the set A. The sharp transition between membership and non-membership is similar to that observed in Equation 3.2. In practice, is this sharp boundary between violators and non-violators realistic? Should there be a traffic summons issued to drivers who are caught at *70.5* km/h? or at *70.8* km/h? In set-theoretical terms, does the person driving at *70.5* km/h or *70.8* km/h belong to the set of violators? In reality, traffic officers are aware of the imprecision of their speed-measuring instrument and are also likely to make allowances for slight infractions, thus they probably set the limit to about *75* km/h before charging a driver. This example points out quite clearly the mismatch between bivalent theory and the multivalent nature of practical life. A much more realistic way to define the membership function would be something like, for example:

$$\begin{aligned} x_1 &= 60 & \mu_A(x_1) &= 0.0 \\ x_2 &= 70.0 & \mu_A(x_2) &= 0.45 \\ x_3 &= 76.0 & \mu_A(x_3) &= 1.0 \end{aligned} \tag{3.3}$$

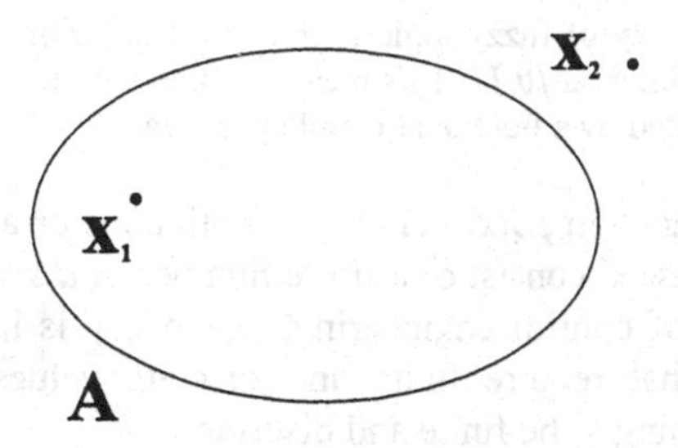

Figure 3.1 Crisp set membership

This is also shown in Fig 3.3 where the transition between the membership and non-membership of individual vehicles in the set of violators is *gradual rather than abrupt* . One might, for example, assign the degree of membership $\mu_A(x) = 1.0$ for drivers that definitely violate the law, whereas those with $\mu_A(x) = 70.8$ are violators only to the degree of about $\mu_A(x) = 0.5$, hence they are not given a summons.

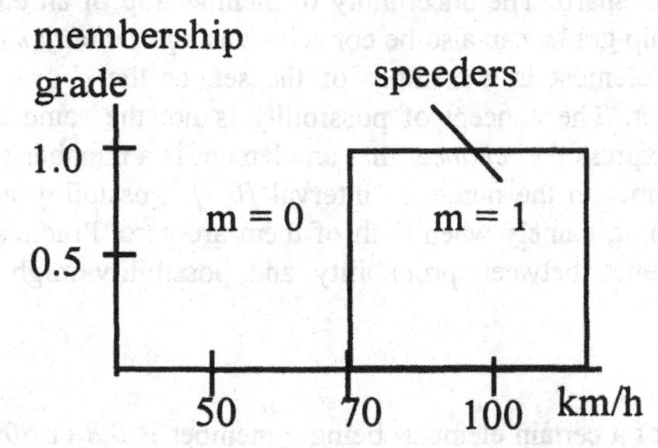

Figure 3.2. Crisp set membership: speed limit violation example.

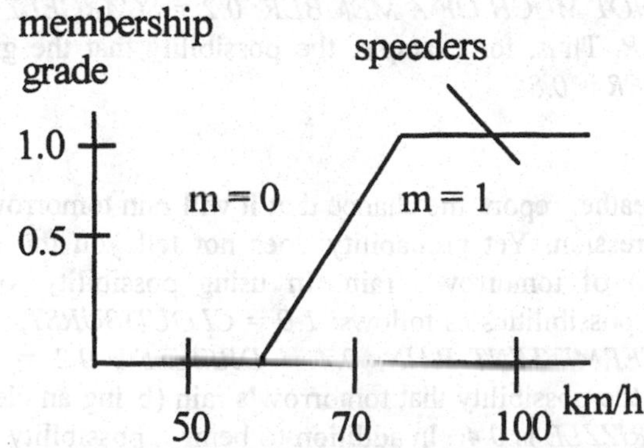

Figure 3.3. Fuzzy set membership: speed limit violation example

The fundamental property of fuzzy logic is that a membership function $\mu_A(x)$ has all of its values in the numeric interval *[0,1]*. This means that an element may be a member of a set to a certain degree indicated by a fractional membership value.

The membership function $\mu_A(x)$ may be a continuous or a discrete function. In this book, the sets discussed consist of a finite number of *discrete* elements. The reason is that in the field of control engineering, set theory is implemented by means of digital computers that require finite and discrete values. Thus the membership functions are also going to be finite and discrete.

3.3 Probability and possibility

Fuzzy set theory is based on the fact that real-world sets have no precise boundaries. A fuzzy set is thus an imprecise and undefined bundle of elements where the transition of from non-membership to membership is gradual and not sudden. Fuzziness implies imprecision, uncertainty, qualitativeness. Fuzzy set theory provides a mathematical method for manipulating sets whose boundaries are imprecise rather than sharp. The uncertainty of membership of an element, i.e. the fractional membership grade, can also be conceived as a *possibility measure*, i.e. the *possibility* that the element is a member of the set, or the *degree* to which the element is a member. The concept of possibility is not the same as *probability*. Probability would express the *chance* that an element is a member and this is also expressed by a number in the numeric interval *[0,1]*. Possibility and probability meet only at one point, namely when both of them are zero. Practical examples to illustrate the difference between probability and possibility might be stated as follows:

Example 3.1:
Let the probability of a certain element being a member is *0.8* or *80%*. Possibility, on the other hand, would express the degree to which the element is a member. To illustrate this point, let us make a linguistic scale of memberships:
1.0 = MEMBER; 0.8 = ALMOST A MEMBER.; 0.6 = MORE-OR-LESS A MEMBER; 0.4 = NOT MUCH OF A MEMBER; 0.2 = SCARCELY A MEMBER; 0 = NOT A MEMBER. Thus, for example, the possibility that the given element is *ALMOST A MEMBER* is *0.8*.

Example 3.2:
According to the weather report, the chance that it will rain tomorrow is *0.8* .This is a probabilistic expression. Yet probability does not tell you the degree (i.e. the quality or strength) of tomorrow's rain. In using possibility, one would first construct a scale of possibilities as follows: *1.0 = CLOUDBURST; 0.8 = RAINING HARD; 0.6 = INTERMITTENT RAIN; 0.4 = DRIZZLE ; 0.2 = SLIGHT RAIN.* Thus, for example, the possibility that tomorrow's rain (being an element of the set *RAIN*) will be a *DRIZZLE* is *0.4*. In addition to being a possibility measure of *0.4*, the linguistic expression *DRIZZLE* also represents the membership value of the element (i.e. tomorrow's rain) in the set *RAIN*.

3.4 Subset

Let E be a set and let A be a subset of E: $A \subset E$. Thus each element of E is also an element of A.

Example 3.3:

$E = \{ x_1,x_2,x_3,x_4,x_5,x_6\}$; $A = \{x_3,x_4,x_5\}$; $A \subset E$; $x \in E$; determine the membership function for subset A with reference to each element of set E :

$$\mu_A(x_1) = 0;\ \mu_A(x_2) = 0;\ \mu_A(x_3) = 1;\mu_A(x_4) = 1;\ \mu_A(x_5) = 1;\ \mu_A(x_6) = 0;$$

A membership value μ_A (x) of an element $x \in A$ *in reference to universe of discourse* E where $A \subset E$ means the participation of an element of E in set A. According to convention, set A can be written in a form where in every argument of the set each element is shown with its corresponding membership value:

$$A = \{ \{x_1,0),\ (x_2,\ 0),\ (x_3,\ 1),\ (x_4,1),\ (x_5,1),\ (x_6,0)\} \tag{3.4}$$

Note that the curly bracket is used as a general symbol for sets.

3.5 Ordered pairs

In a set with an argument of ordered pairs, such as Equation 3.4, the argument consists of two elements, a and b, in a prescribed order. The ordered pair a and b, comprises the argument (a, b) where a is the first and b is the second coordinate. There is no specific relationship or rule applicable to a and b to form an ordered pair. The only restriction is that the elements of such a pair be in a certain order, such as that a be first and b be second *consistently*.

3.6 Definition of a set

A rigorous definition of a set follows: let E be a set and let x be an element of E. Then a subset A of E is a set of ordered pairs

$$\{ [x,\ \mu_A(x)] \},\ \forall x \in E \tag{3.5}$$

where $\mu_A(x)$ is the degree or grade or membership of x in A.
If $\mu_A(x)$ takes its values in a set M called the *membership set*, one may say that x takes its values in M through the function $\mu_A(x)$:

$$x \underset{\mu_A(x)}{\rightarrow} M \tag{3.6}$$

This function is called a *membership function.* If $M = \{0,1\}$, the subset A is understood as a crisp set and the functions $\mu_A(x)$ will be *Boolean* (binary) functions. If $M = [0,1]$, the subset A is understood as a fuzzy set and the functions $\mu_A(x)$ will be fuzzy functions. The element symbols may be omitted for clarity, thus the set can be written in terms of its membership values only as a so-called *membership vector*:

$$A = (0,0,1,1,1,0) \tag{3.7}$$

In a general form:

$$A = \{ \mu_1 (x), \mu_2(x) , \dots, \mu_n(x)\} \tag{3.8}$$

<u>Definition</u>: A set is completely defined by its membership vector. That is, to determine a set it is sufficient to calculate the individual elements of its membership vector.

3.7 Empty set

An empty set is defined as a set without any members, indicated by the symbol □.

3.8 Universal set

The universal set is the set of all sets, indicated by the symbol **E**.

3.9 Universe of discourse

Let us again define *E* as an instrument scale with six elements and a subset *A* with three elements which are also members of *E*. Then *E* is called the ***universe of discourse*** (or space, or range) of *A*.[1]

<u>Example 3.4:</u> Let there be *k* discrete inputs *u[k]* of a single-input control system which are members of an input set *U* = *{u[k]}*. *U* is therefore the universe of discourse of the input, i.e. it contains all possible values of *u[k]* that can occur.

3.10 Bivalence

Bivalence means two-valuedness: something is either true or not true, either black or white, either zero or one. The classical logic of *Aristotle* made bivalence one of the cornerstones of Western culture. We expect a statement to have only one of two values: *true* or *false*. There is nothing in between, any middle value is excluded. Bivalence is deeply embedded into our thinking, our historical tradition, and even our ethics. It seems to be a "law of thought" which we find "self-explanatory". For example, someone is either my friend or my enemy. The usual consequences of such ethics may be quite harsh. We have little tolerance for the excluded middle: we use derogatory terms for someone who tells "half-truths". Computer science is based on bivalence. *Boolean* mathematics is said to embody the laws of truth in mathematical language. In our science and technology truth and accuracy have become interrelated and considered as an indispensable part of the scientific method: if something is not absolutely accurate, then it is not true. Two times two is four, absolutely. In fact, we used to laugh derisively at the first primitive pocket calculators that gave *3.9999999* as a result.

3.11 Multivalence

There exists, however, a considerable mismatch between the real world and ourbivalent view of it. The real world contains an infinite number of shades of gray be tween the black and white extremes. For example, think of medical diagnosis or a legal decision, where the practitioner must integrate in his mind a number of different, often contradictory factors and arrive at an intuitive judgment representing a "gray-scale" description of the degree of a patient's illness or the degree of guilt of an accused. It seems that in the real world *everything is a matter of degree*. The real world is not bivalent, it is multivalent with an infinite spectrum of options instead of just two. In technical terms, the real world is analog, not digital, with infinite shades of gray between black and white. Absolute truth and accuracy exists only in the artificial world of mathematical logic, which remain black and white as extreme cases of gray. Insisting on the black and white extremes as being only valid is a cultural bias, nothing else. The objective of fuzzy logic is to capture these shades of gray, these degrees of truth. Fuzzy logic deals with *uncertainty* and the partial truth represented by the various shades of gray in a systematic and rigorous manner.

Human communication contains much uncertainty in the form of vague, imprecise, inaccurate, verbal expressions. We use the same words but may attribute different meanings to them. To humans, *words represent not one idea but a set of ideas*: just think of words like: house, school, car. We use intuitive value judgments to describe as *to what degree* the house, school, or car fit into the set. Perhaps in our personal view a shack is a house to the degree of *5* %, a school building is a school to the degree of *50* %, and a Porsche is a car to the degree of *100*%. The sets "house", "school", and "car" are *fuzzy sets*. Humans can reason with fuzzy sets. Computers can only "think" in terms of bivalent truths: *0*s and *1*s. They cannot understand the fuzzy terms of human communication. However, fuzzy logic can translate degrees of truth into a form that computers can process. Fuzzy logic is a multivalued rather than a two-valued logic that can make computers reason with fuzzy sets like humans.

3.12 Fuzzy numbers

Consider a number as a spike drawn with a chalk on a blackboard at, say, *0* on a number line, as shown in Figure 3.4, with a height of, say, unity. This is a so-called *crisp* (i.e. non-fuzzy) number. A crisp number can also be considered as a crisp *set*, with members either inside the set or outside of it. Let us call the spike set arbitrarily *ZERO*, thus every number is either in the *ZERO* set or out of it. Assume now that we use a sponge to smear the chalked line left and right. This smeared version represents the numbers in the neighborhood of the set *ZERO*. One can call this set, smeared to the left and to the right, shown as a triangle, in Figure 3.5, arbitrarily *ALMOST ZERO*. One can observe that the set *ZERO* has the mathematical abstraction of zero area, while the set *ALMOST ZERO* has a finite area. It is possible to draw a triangular set, *NEAR ZERO*, which is even more smeared to the left and to the right, i.e. has a greater area, than the set *ALMOST ZERO*. One could intuitively state, that the set *ALMOST ZERO* is fuzzy and the set *NEAR ZERO* (Figure 3.6) is "fuzzier" than the set *ALMOST ZERO* [3]. The terms "fuzzy numbers" and "fuzzy sets" are interchangeable. Clearly, *crisp numbers represent a special case of fuzzy numbers.*

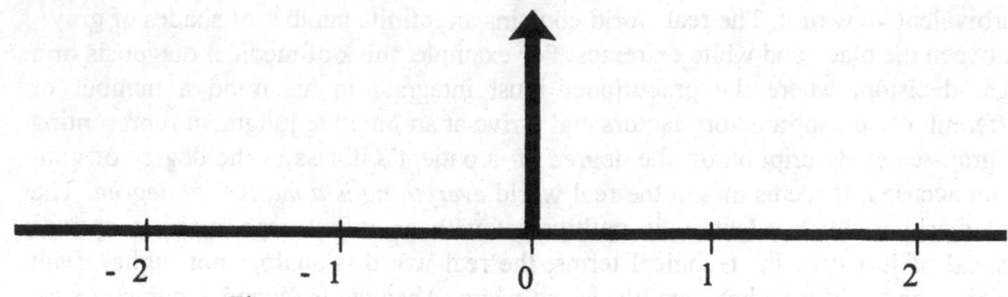

Figure 3.4. Representation of crisp number "zero"

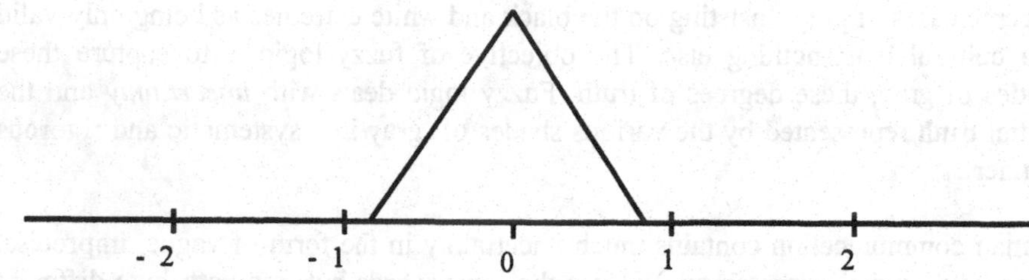

Figure 3.5. Representation of fuzzy number "almost zero"

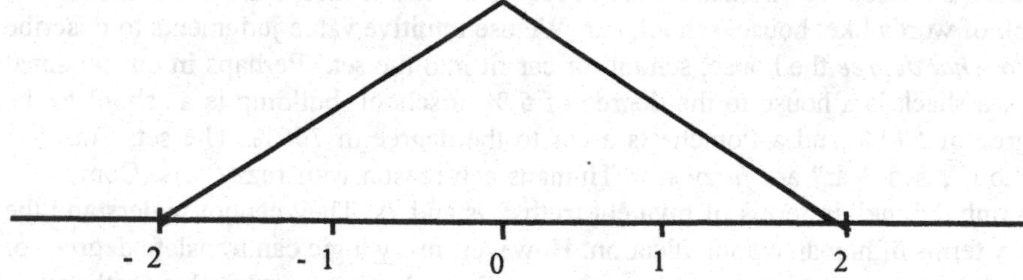

Figure 3.6. Representation of fuzzy number "near zero"

3.13 Fuzzy numbers: basic requirements

Figure 3.6 illustrated a graphic representation of a typical fuzzy number X which was shown as a fuzzy set with a finite area plotted along the number axis, i.e. a finite continuous universe of discourse U. As regards the form of X, there are certain requirements as follows:

- X must be *normal*: $$max \ \mu_X(u) = 1, \ u \in U \quad (3.9)$$

This means that at any element $u \in U$, the maximum height of the membership function must be *1*.

- X must be *convex*: $$\mu_X [\lambda u_1 + (1 - \lambda) u_2] \geq min [\mu_X (u_1), \mu_X (u_2)] \quad (3.10)$$
for $u_1, u_2 \in U, \ \lambda \in [0,1]$

Convexity carries information about the interior connectivity and shape of the fuzzy number. In other words, it prevents "holes" and "bays" within the boundary.. Refer to Figure 3.7. This is to exclude the occurrence of more than one fuzzy value that pertains to any specific crisp abscissa value within the same membership function..

An alternate formulation of fuzzy set convexity is that whenever the membership values are monotonically increasing, or monotonically decreasing, or increasing then decreasing when traversing the universe of discourse from its minimum to its maximum values. That is, if for any element $x, y, z \in A^f$ then the expression $x < y < z$ implies that $\mu_A^f \geq min\ [\mu_A^f(x), \mu_A^f(z)]$, then A^f is a convex fuzzy set. (If the membership values are first monotically decreasing then increasing, then the fuzzy set is not convex, the decrease followed by an increase may be construed as a "bay"). A special property of two convex fuzzy sets is that their intersection is also convex.

3.14 Linguistic variables

A linguistic variable *u* in the universe of discourse *U* is defined in terms of the set names of linguistic values *T(u)*, with each value being a fuzzy number defined on *U*. For example, if *u* means speed, then its term set *T(u)* could be:

$$T(speed) = \{slow, medium, fast\}$$

over the universe of discourse $U = [0,100]$. *Slow, medium, fast* are linguistic values of the linguistic variable *speed.*

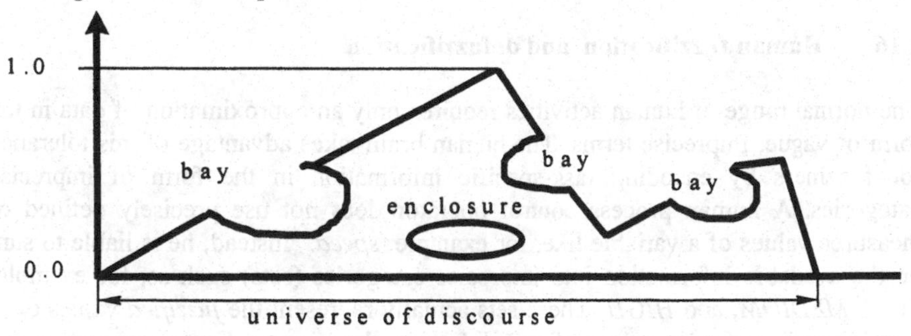

Figure 3.7. Fuzzy numbers or fuzzy sets: basic requirements

A linguistic variable *u* in the universe of discourse *U* is defined in terms of the set names of linguistic values *T(u)*, with each value being a fuzzy number defined on *U*. For example, if *u* means speed, then its term set *T(u)* could be:

$$T(speed) = \{slow, medium, fast\}$$

over the universe of discourse $U = [0,100]$. *Slow, medium, fast* are linguistic values of the linguistic variable *speed.*

3.15 Logical implication and rules of inference

Fuzzy logic means *reasoning* with fuzzy numbers and fuzzy sets. There is another aspect of human thinking: *logical implication,* (an alternate expression indicating reasoning), which consists of the formulation of a connection between cause and effect, or a *condition* and its *consequence.* In our technical careers, such logical implications are followed virtually in every situation, such as, for example, when we operate a machine, solve mathematical problems, program a computer, follow a procedure in an instruction manual, or make a decision as to which instrument to purchase. In such cases, we consciously or unconsciously follow so-called *rules of inference* which have the form:

$$\textbf{IF}\ \text{cause}_1 = A\ \text{ and cause}_2 = B\ \textbf{ THEN}\ \text{effect} = C\,.$$

where *A,B* and *C* are sets. These statements may be considered as rules-of-thumb, like in the following practical situation:

IF speed is *HIGH* **THEN** throttle should be *REDUCED*

where terms *HIGH* and *REDUCED* represent *fuzzy sets.HIGH* is a function defining the speed of motion, while *REDUCED* is a function defining degrees of throttle position.

The intelligence lies in associating these two fuzzy terms by means of a *fuzzy inference* expressed in heuristic **IF...THEN** terms.

3.16 Human fuzzification and defuzzification

The normal range of human activities requires only an approximation of data in the form of vague, imprecise terms. The human brain takes advantage of this tolerance for fuzziness by encoding task-specific information in the form of imprecise categories. A human process control operator does not use precisely defined or measured values of a variable like, for example, *speed.* Instead, he is liable to sum up the available information into imprecise categories (sets) such as, for example, *LOW, MEDIUM,* and *HIGH.* These sets actually represent the *fuzzified* values of a certain number of crisp values of *speed.* In turn, the operator formulates and carries out his control strategy on the basis of the fuzzified values of every input and output variable. Thus the data stream that he observes is reduced to just what is necessary to carry out the given task with a sufficient accuracy. In turn, the human operator processes the now fuzzy quantities and arrives at a fuzzy variable as his action output. Similar fuzzy sets are used in industrial plant/process control and the output, represented by a fuzzy set, must again be converted into a single crisp output value by means of *defuzzification.* This is necessary because most industrial applications require an unequivocal, crisp, i.e. non-fuzzy action output, and this defuzzified value is used, for example, to set a lever or switch to a specific crisp position[2] .Examples of human defuzzification include the ability to decipher sloppy handwriting, to understand broken or accented language, or to recognize someone after a long absence (even if he shaved off his beard and lost weight, or she changed her hairdo and wears glasses). Defuzzification in this sense is akin to feature extraction in pattern recognition. It seems that the ability to manipulate fuzzy sets is one of the human brain's most important activities. In as much as there is no mathematical model to follow, fuzzy decision processes have a minimal computational overhead. Perhaps this human ability developed throughout the ages, because the process of trading off accuracy for processing speed is more conducive to biological survival in life-critical situations than equivalent more precise models. Besides, in complex systems mathematical accuracy loses its meaning. The *Principle of Incompatibility* of Zadeh [1973] states: "As the complexity of a system increases, our ability to make precise and significant statements about its behavior diminishes until a threshold is reached, beyond which precision and significance (or relevance) become almost mutually exclusive characteristics."

3.17 A qualitative summary of fuzzy logic

- Classical logic is *bivalent*, that is, it recognizes only two values: true or false. In contrast, fuzzy logic is *multivalent*, that is, it recognizes a multitude of values: it holds that *truth is a matter of degree* and defines such a degree by assigning a truth value in the numeric interval *[0,1]*.
- Fuzzy logic is a way of *handling uncertainty* by expressing it in terms of a degree of certainty in the numeric interval *[0,1]*, where certainty is assigned the value of *1*.
- Imprecise, qualitative, verbal, fuzzy expressions inherent in human communication also possess various degrees of uncertainty and are thus amenable to handling by fuzzy logic.
- In the human reasoning process consisting of *logical implications* or logical inference, both the *conditional* (i.e. input) and the *consequent* (i.e. output) variables are assigned truth values in the numeric interval *[0,1]*. Thus the fuzzy rules are expressions of the inherently fuzzy human reasoning process.
- Fuzzy logic can systematically translate the inherently fuzzy terms of human communication into numeric values understood by computers. Since computers are general purpose machines that interface with physical, chemical, thermal and biological processes, the inherently fuzzy terms of human communication can be used directly to exchange information between humans and such processes in a rigorous and systematic manner.

The purpose of this work is to provide the methodology for such an information exchange.

Notes

1. Strictly speaking, a universe of discourse is always a crisp set, thus every fuzzy set should properly be called a fuzzy subset. We will, however, not adhere to this strict rule in order to simplify our expressions.
2. There are some interesting exceptions in the food industry. For example, in evaluating the quality of cheese, the system output must be left in fuzzy categories (i.e. without defuzzification) each of which corresponds to the choices of experienced human tasters.
3. The degree of fuzziness is measured by means of the fuzzy entropy theorem given an elegant geometric definition by Kosko in the reference work listed below.

References

ABDELNOUR GM et al:"Design Of A Fuzzy Controller Using Input and Output Mapping Factors."IEEE Trans.Sys.,Man,Cybern.,1991;21;5;952-960.

KANDEL A., LEE , SC: *Fuzzy Switching and Automata*. John Wiley and Sons, New York 1976.

KOSKO B:*Neural Networks and Fuzzy Systems*. Prentice Hall, 1992.

LEE CC: "Fuzzy Logic In Control Systems." IEEE Transcations on Sys., Man,Cybern.,1990;20;3;404-435;

MAMDANI EH, ASSILIAN S:"An Experiment In Linguistic Synthesis With A Fuzzy Controller." Int.Journal Man-Machine Studies, 1975;7;1-13;

MENDEL JM: "Fuzzy Logic Systems for Engineering: A Tutorial." Proc.IEEE, 1995;83;3;343-377.

NEGOITA, CV., RALESCU, DA:"Fuzzy Systems and Artificial Intelligence." Kybernetes, 1974; 3.

RAGADE, RK., GUPTA, MM:"Fuzzy Set Theory: An Introduction." In *Fuzzy Automata and Decision Processes*, Gupta MM, Saridis GN, Gaines BR, eds. North Holland, 1976.

SCHWARTZ D, KLIR G, LEWIS HW, EZAWA Y: "Applications of Fuzzy Sets and Approximate Reasoning". Proc.IEEE 1994;82;4.

ZADEH , LA: "Outline Of a New Approach To the Analysis of Complex Systems and Decision Processes." IEEE Trans. on Sys,Man, Cybern.,1973; 3: 28-44

ZADEH, LA: "Towards a Theory of Fuzzy Systems. In *Aspects of Network and Systems Theory,* Kalman RE and DeClaris N.eds.New York, Holt,Rhinehart and Winston 1971.

ZADEH, LA:"A Fuzzy Algorithmic Approach to the Definition of Complex or Imprecise Concepts. " Int. Journal of Man-Machine Studies, 1976; 8.

YASUNOBU S, MIYAMOTO S:"Automatic Train Operation System by Predictive Fuzzy Control." In: *Industrial Applications of Fuzzy Control*, M.Sugeno Ed.Amsterdam, The Netherlands, Elsevier 1985;1-18;

4 SET OPERATIONS

4.1 Set operations in general

Control theory investigates the interaction between time-variant energy and physical systems. Interaction implies various operations to be carried out by means of certain rules. If the energy (mechanical, electrical, thermal, chemical, etc.) as well as the systems (or their mathematical models) are expressed in terms of sets, one must first determine the rules of interaction between sets in general. In as much as the rules of interaction affect the membership vectors of the respective sets, one must first establish the rules of combining membership vectors in general.

4.2 Set operations in the same universe of discourse.

The universes of discourse of a plant or process input, U, and the output, Y, contain all sets of input signals $u[k] \in U$ and output signals $y[k] \in Y$ on all input and output terminals respectively, without any regard to their timing. In these cases operations between input sets $u[k]$ occur within the same universe of discourse, U, and operations between output sets $y[k]$ occur within the same universe of discourse, Y.

Although in crisp logic theory the operations between sets are essentially the same as the *Boolean* operations **AND, OR** and **NOT**, in fuzzy logic there are many more operators called *triangular norms* or ***t***-norms and their dual, the ***s***-norms (also called ***t*** co-norms). The possible operations between sets provide a pool of operators to be used in logical implications **IF...THEN**. The designer has thus many more degrees of freedom in choosing the optimum fuzzy implication operator for his specific control task. In this chapter, only a brief outline of the general forms of the ***t*** and ***s*** operators are given.[1] In the following, operations between crisp sets *within the same universe of discourse* and using discrete signal vectors will be defined. This discussion will also serve as a review of *Boolean* algebra.

4.3 Intersection of crisp sets

Given two sets A and B , $A \subset E$, $B \subset E$, where E is their common universe of discourse, the intersection $A \cap B$ is a set of all elements x which are members of both A and B. This is illustrated in the *Venn* diagram of Figure 4.1.

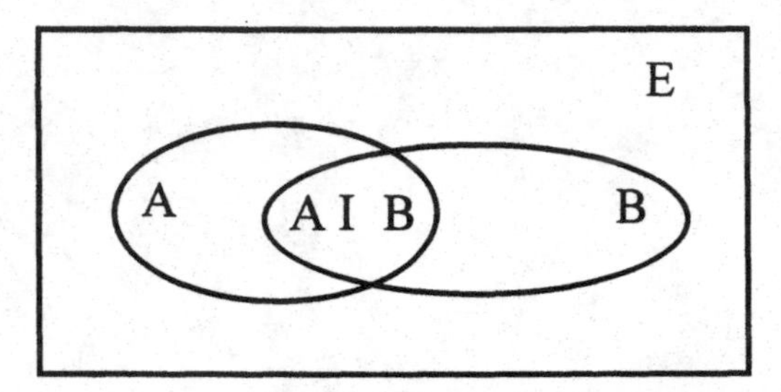

Figure 4.1. Intersection of crisp sets.

Our aim is to determine the membership vector of the intersection $A \cap B$ from the known individual membership vectors of sets A and B. The symbol "$\cap$" is read as "cap".

The membership vector of A: $\mu_A(x) = 1$ if $x \in A$
$\mu_A(x) = 0$ if $x \notin A$

The membership vector of B: $\mu_B(x) = 1$ if $x \in B$
$\mu_B(x) = 0$ if $x \notin B$

The intersection vector contains all elements that are members of both A and B: Thus the intersection's membership vector is:

$$\mu_{A \cap B}(x) = 1 \quad if\ x \in A \cap B$$
$$\mu_{A \cap B}(x) = 0 \quad if\ x \notin A \cap B$$

Thus

$$\mu_{A \cap B}(x) = \mu_A(x) \cdot \mu_B(x) \tag{4.1}$$

where the operator "." signifies the *Boolean* **AND** function which is carried out on each element pair on the basis of the following truth table:

Table 4.1.Truth table for crisp intersection

A	B	A ∩ B	Membership
0	0	0	non-member
0	1	0	non-member
1	0	0	non-member
1	1	1	member

Observe that the intersection is the largest subset of the universe of discourse *E*, which is at the same time part of *A* and also of *B*. The intersection is the *overlapping* portion of sets *A* and *B* and, as a result, it is always *smaller* than any of the individual sets *A* and *B*. For this reason, the membership vector of intersection $A \cap B$ (in other words, the components of the membership vector which are the membership grades) is calculated from the known individual membership vectors of *A* and *B* as follows:

$$\mu_{A \cap B}(x) = min\ [\mu_A(x),\ \mu_B(x)] \tag{4.2}$$

Example 4.1:The intersection's membership vector is calculated as follows:
$E = \{ x_1, x_2, x_3, x_4, x_5 \}$; $A = \{ 0, 1, 1, 0, 1\}$; $B = \{ 1, 0, 1, 0, 1 \}$; $A \subset E$; $B \subset E$;
The intersection as a set is:
$A \cap B = \{ (0 . 1), (1 . 0), (1 . 1), (0 . 0), (1 .1) \}$ where "." is a *Boolean* **AND**
The membership vector:
$\mu_{A \cap B}(x) = \{ 0, 0, 1, 0, 1 \}$

4.4 Disjointed crisp sets

Sets that have no common members are called disjointed sets (Figure 4.2). That is, the intersection of disjointed sets is the empty set.

$$A \cap B = \square$$

4.5 Union of crisp sets

Given sets *A* and *B*, $A \subset E$, $B \subset E$, where *E* is their common universe of discourse, the union $A \cup B$ is a set of all elements *x* which are members of set *A* or set *B* *or* both *A* and *B*. This is illustrated in the *Venn* diagram of Figure 4.3. Our aim is to determine the membership vector of the union $A \cup B$ from the known individual membership vectors of sets *A* and *B*. The symbol "∪" is read as"cup.

The membership vector of *A*: $\mu_A(x) = 1 \quad if\ x \in A$

$$\mu_A(x) = 0 \quad if\ x \notin A$$

The membership vector of B: $\mu_B(x) = 1 \quad if\ x \in B$

$$\mu_B(x) = 0 \quad if\ x \notin B$$

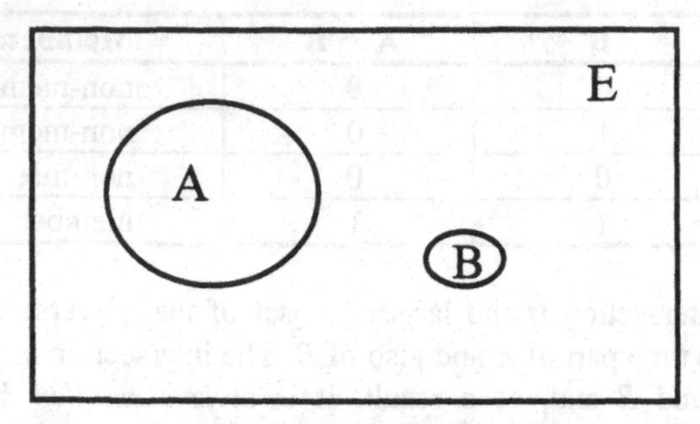

Figure 4.2. Disjointed crisp sets.

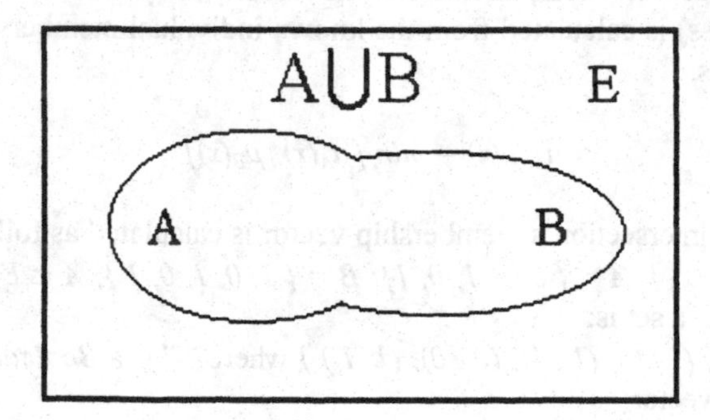

Figure 4.3. Union of crisp sets.

The union vector contains all elements that are members of A or B or both.
Thus the union's membership vector is: $\mu_{A\cup B}(x) = 1 \quad if\ x \in A \cup B$

$$\mu_{A\cup B}(x) = 0 \quad if\ x \notin A \cup B$$

Thus

$$\mu_{A\cup B}(x) = \mu_A(x) + \mu_B(x) \tag{4.3}$$

where the operator "+" signifies the *Boolean* **OR** function which is carried out on each element pair on the basis of the following truth table:

Table 4.2. Truth table for crisp union.

A	B	A ∪ B	Membership
0	0	0	non-member
0	1	1	member
1	0	1	member
1	1	1	member

Observe that the union is the smallest subset of the universe of discourse *E*, which includes both sets *A* and *B*. The union is the *contour* that includes both sets *A* and *B* and, as a result, it is always *larger* than any of the individual sets *A* and *B*. For this reason, the membership vector of union $A \cup B$ (in other words, the components of the membership vector which are the membership grades) is calculated from the known individual membership vectors of *A* and *B* as follows:

$$\mu_{A \cup B}(x) = max\ [\mu_A(x),\ \mu_B(x)] \tag{4.4}$$

The following example illustrates the calculation of the union's membership vector.

Example 4.2:
$E = \{ x_1, x_2, x_3, x_4, x_5 \}$; $A = \{ 0, 1, 1, 0, 1\}$; $B = \{ 1, 0, 1, 0, 1 \}$; $A \subset E$; $B \subset E$;
The union as a set is
$A \cup B = \{ (0 + 1), (1+ 0), (1+ 1), (0+ 0), (1 + 1) \}$ where "+" is a *Boolean* **OR.**
The membership vector:
$\mu_{A \cup B}(x) = \{1, 1, 1, 0, 1 \}$

4.6 Complement of crisp sets

Let *A* be a subset of universe of discourse *E*. The complement of *A* with respect to *E*, is *A'*, a set of all elements $x \in E$ which are not members of *A*. This is shown in the *Venn* diagram of Figure 4.4. If $x \in A$ and $x \notin A'$, then the membership functions of *A* and its complement *A'* are:

$$\mu_A(x) = 1 \text{ and } \mu_{A'}(x) = 0$$

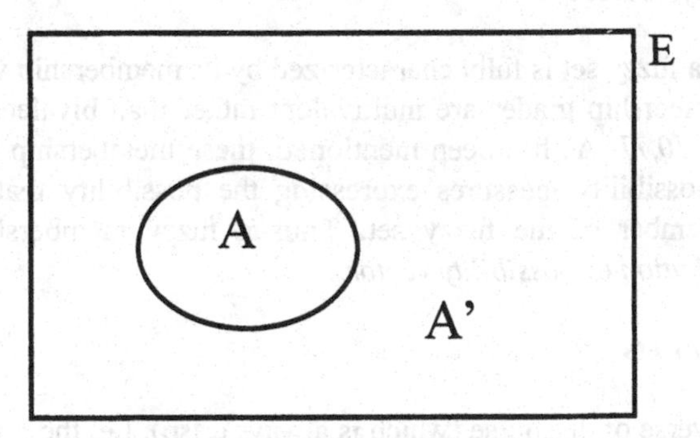

Figure 4.4. Complement of crisp sets.

The complement's membership vector is calculated as follows:

$$\mu_A'(x) = 1 - \mu_A(x) \tag{4.5}$$

The following example illustrates the calculation of the complement's membership vector.

Example 4.3:
$E = \{ x_1, x_2, x_3, x_4, x_5 \}$; $A = \{ 0, 1, 1, 0, 1\}$;
The complement set of A with respect to E is: $A' = \{ 0', 1', 1', 0', 1'\}$
where the apostrophe signifies *Boolean* **NOT**.
The membership vector of the complement of A is:
$\mu_A'(x) = \{ 1, 0, 0, 1, 0\}$

4.7 Fundamental properties of crisp sets

The fundamental properties of crisp sets lie in the definitions of classical bivalent logic. Assume that the crisp set A is a subset of universe of discourse E. Figure 4.4 shows that set A and its complement, A', are by implication always *disjointed,* that is, their *intersection* is always the empty set:

$$A \cap A' = \Box \tag{4.6}$$

It is obvious that A and A' do not overlap, that is, any middle value between A and A' is excluded. At the same time, the *union* of a crisp set and its complement is their universe of discourse. In this case, A and A' completely fill the universe of discourse E, there is nothing left over, there is no "underlap":

$$A \cup A' = E \tag{4.7}$$

4.8 Possibility vectors

Like a crisp set, a fuzzy set is fully characterized by its membership vector, except in this case its membership grades are multivalent rather than bivalent and are in the numeric interval $[0,1]$. As has been mentioned, these membership grades can also be considered possibility measures expressing the possibility that the respective element is a member of the fuzzy set. Thus a fuzzy membership vector is a *possibility distribution* or *possibility vector.*

4.9 Fuzzy subsets

Let E be the universe of discourse (which is always crisp). Let the possibility vectors be $A \subset E$, $B \subset E$, and $M = [0,1]$, where M is the set of the membership values of all sets in the example.

Then $A \subset B$ for all element x if

$$\mu_A(x) \leq \mu_B(x) \tag{4.8}$$

In other words, A is contained in B if the membership value of each element $x \in E$ in A is less then or equal to the corresponding membership value $x \in E$ of B. That is, all elements of E that fit into A also fit into B, with space still left over in some cases.

Example 4.4:
$E = \{ x_1, x_2, x_3, x_4 \}; M = [0,1] ; A = \{ 0.4, 0.2, 0, 1\}; B = \{ 0.3, 0, 0, 0 \}$
then $B \subset A$ because $0.3 < 0.4 ; 0 < 0.2; 0 = 0; 0 < 1;$

In the following, operations between fuzzy sets represented by discrete vectors will be defined.

4.10 Intersection of fuzzy sets

Let E be the universe of discourse and $x \in E$, $M = [0,1]$; let $A \subset E$, $B \subset E$, then the intersection $A \cap B$ is the largest subset of the universe of discourse E, which is at the same time part of A and also of B. The intersection is the *overlapping* portion of sets A and B and, as a result, it is always *smaller* than any of the individual sets A and B. For this reason, the membership vector of intersection $A \cap B$ (in other words, the components of the membership vector which are the membership grades) is calculated from the known individual membership vectors of A and B as follows:

$$\mu_{A \cap B}(x) = \min \left[\mu_A(x), \mu_B(x)\right] \tag{4.9}$$

This is the same as Equation 4.2 which applied to crisp intersection, except that now $M = [0,1]$ whereas under the conditions of Equation 4.2 it was $M = \{0,1\}$.

Example 4.5:
$E = \{ x_1, x_2, x_3, x_4, x_5 \}; M = [0,1];$
$A = \{ 0.2, 0.7, 1, 0, 0.5 \}; B = \{ 0.5, 0.3, 1, 0.1, 0.5 \};$
then $A \cap B = \{0.2, 0.3, 1, 0, 0.5\}$

A fuzzy **AND** function can be defined as follows: Let the universe of discourse be $E(x)$ with two fuzzy subsets, $x \in A$, $x \in B$. Then

$$\underset{\mu_A}{x \in A} \ AND \ \underset{\mu_B}{x \in B} \rightarrow \underset{\mu_{A \cap B}}{x \in A \cap B} \text{ for } \forall x$$

where $\forall x$ means *"for all x"*. As regards graphical representations, *Venn* diagrams are insufficient for fuzzy sets because they do not contain any information about membership values other than 0 and 1. The diagrams shown for fuzzy sets use a scale of 0 to E, the universe of discourse, on the horizontal axis, and a scale of $[0,1]$ on the vertical axis. Figure 4.5 illustrates the graphical representation of the intersection of fuzzy sets.

4.11 General fuzzy intersection operators

The membership function $\mu_{A \cap B}(u)$, $u \in U$, of the intersection $A \cap B$ is defined point-by-point by

$$\mu_{A \cap B}(u) = \mu_A(u)\ t\ \mu_B(u) \leq \min [\ \mu_A(u),\ \mu_B(u)]$$

where ***t*** is the *triangular norm* or ***t****-norm* of a generalized intersection. The ***t***-norm is a two-place function, $t\colon [0,1] \ X\ [0,1] \rightarrow [0,1]$ where '*X*' denotes the *Cartesian product* operator (see Section 4.20). The ***t***-operator indicates a kind of mapping between fuzzy membership functions, each of which is in the interval *[0,1]*. To understand this graphically, one can assume that each of these fuzzy sets in *[0,1]* on the left-hand side of the arrow is a membership function of a certain shape and the ***t***-operation is carried out between them according to its specification, point-by-point. Such a mapping function must satisfy certain conditions, such as:

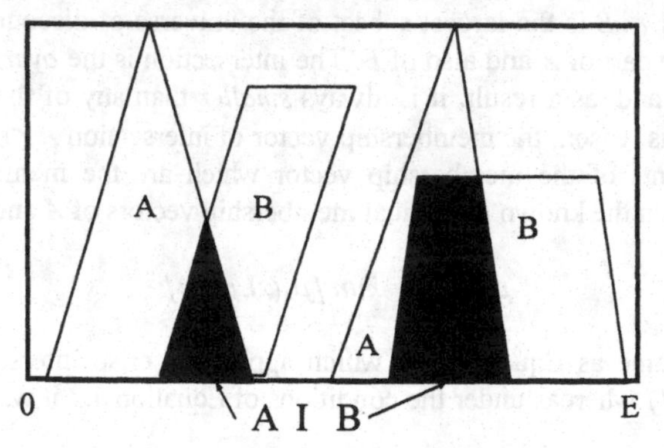

Figure 4.5. Intersection of fuzzy sets.

Boundary conditions: $x\ t\ 0 = 0,\ \forall u \in [0,1]$

This can be readily seen from the *Venn* diagram of an intersection, where the intersection (i.e. common area) of any set with the empty set is obviously the empty set.

$$x\ t\ 1 = x,\ \forall u \in [0,1]$$

Likewise, a *Venn* diagram would immediately show that the intersection of any set with the universal set must be the set itself.

Commutative condition: $x\ t\ y = y\ t\ x$

Associative condition: $x\ t\ (y\ t\ z) = (x\ t\ y)\ t\ z$

Non-decreasing condition: for $x \leq y$ *and* $w \leq z$, $x\ t\ w \leq y\ t\ z$

The ***t***-norms include: intersection, algebraic product, logarithmic product, inverse product, bounded product and drastic product. Some ***t***-norms for $x, y \in [0,1]$ are:

Intersection: $x\,t\,y = min\,(x,y)$

Algebraic product $x\,t\,y = xy$

Drastic product: $x\,t\,y = \begin{cases} x \text{ when } y = 1 \\ y \text{ when } x = 1 \\ 0 \text{ when } x,y < 1 \end{cases}$

Example 4.6:
As an example, the intersection and the drastic product will be plotted. Assume two membership functions x and y whose values are in $[0,1]$ are drawn with solid lines over a given universe of discourse, as in Figure 16. The intersection (***min***) is the largest common area of the two functions, shown with a dashed line. The drastic product (***dras***) is drawn with the dotted line, as follows: when $y = 1$, then ***dras*** is x is itself, when $y < 1$, then ***dras*** becomes zero; when $x = 1$, then ***dras*** becomes y itself, when $x < 1$ then ***dras*** becomes zero again, when $y = 1$ then ***dras*** becomes x itself again.

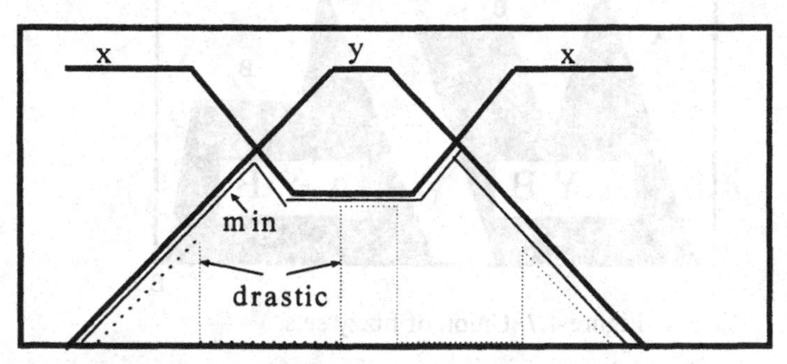

Figure 4.6. *t*-norms: intersection and drastic product

4.12 Union of fuzzy sets

Let E be the universe of discourse, $x \in E$, $M = [0,1]$; fuzzy sets $A \subset E$; $B \subset E$. Then the union $A \cup B$ is the *smallest* subset of the universe of discourse E, which includes both fuzzy sets A and B. The union is the *contour* that includes both fuzzy sets A and B and, as a result, it is always *larger* than any of the individual sets A and B. For this reason, the membership vector of union $A \cup B$ (in other words, the components of the membership vector which are the membership grades) is calculated from the known individual membership vectors of A and B as follows:

$$\mu_{A \cup B}(x) = max\,[\mu_A(x), \mu_B(x)] \tag{4.10}$$

This is the same as Equation 4.4 given for crisp sets, except that here $M = [0,1]$. The following example illustrates the calculation of the union's membership vector.

Example 4.7:
$E = \{ x_1, x_2, x_3, x_4, x_5 \}$; $M = [0,1]$; $A = \{ 0.2, 0.7, 1, 0, 0.5 \}$; $B = \{ 0.5, 0.3, 1, 0.1, 0.5 \}$; then $A \cup B = \{0.5, 0.7, 1, 0.1, 0.5\}$

A fuzzy **OR** function can be defined as follows: let the universe of discourse be $E(x)$ with two fuzzy subsets, $x \in A$, $x \in B$. Then

$$\underset{\mu_A}{x \in A} \ OR \ \underset{\mu_B}{x \in B} \rightarrow \underset{\mu_{A \cup B}}{x \in A \cup B} \text{ for } \forall x$$

Figure 4.7 illustrates the graphical representation of the union of fuzzy sets.

4.13 General fuzzy union operators

The membership function $\mu_{A \cup B}(u)$, $u \in U$, of the union $A \cup B$ is defined point-by-point by

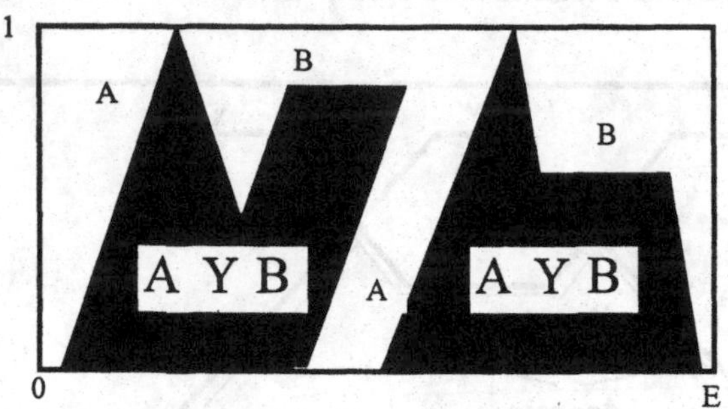

Figure 4.7. Union of fuzzy sets.

$$\mu_{A \cup B}(u) = \mu_A(u) \ s \ \mu_B(u) \leq max \ [\ \mu_A(u), \ \mu_B(u)$$

where ***s*** *is the triangular co-norm* of a generalized union. The ***s*** norm is a two-place function, ***s***: $[0,1] \ X \ [0,1] \rightarrow [0,1]$ where '*X*' denotes the *Cartesian product* operator (see Section 4.20). The ***s*** -operator indicates a kind of mapping between fuzzy membership functions, each of which is in the interval $[0,1]$. To understand this graphically, one can assume that each of these fuzzy sets in $[0,1]$ on the left-hand side of the arrow is a membership function of a certain shape and the ***s***-operation is carried out between them according to its specification, point-by-point. Such a mapping function must satisfy certain conditions, such as:

Boundary conditions: $u \ s \ 0 = u \, , \ \forall u \in [0,1]$

This can be readily verified from the *Venn* diagram of a union, where the union (i.e. common contour) of any set with the empty set is obviously the set itself.

$$u \ s \ 1 = 1, \ \ \forall u \in [0,1]$$

Likewise, a *Venn* diagram would immediately show that the union of any set with the universal set must be the universal set itself.

Commutative condition: $x\ s\ y = y\ t\ x$
Associative condition: $x\ s\ (y\ s\ z) = (x\ s\ y) s\ z$
Non- decreasing condition: for $x \leq y$ *and* $w \leq z$, $x\ s\ w \leq y\ s\ z$

The *s*-norms include: union, algebraic sum, bounded sum, logarithmic sum, disjoint sum, and drastic sum. Some *s*-norms for $x, y \in [0,1]$ are:

Union: $x\ s\ y = max\ (x,y)$,
Algebraic sum: $x\ s\ y = x + y - xy$

4.14 Complement of a fuzzy set

Let E be the universe of discourse, $x \in E$, $M = [0,1]$; fuzzy set $A \subset E$;
The complement of A with respect to E, is A', a set of all elements $x \in E$ which are not members of A. This is shown in the diagram of Figure 4.8.

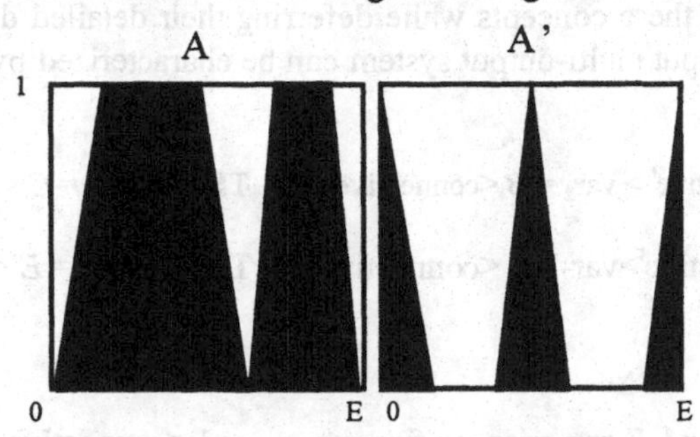

Figure 4.8. Fuzzy complement.

The complement's membership vector is calculated as follows:

$$\mu_A'(x) = 1 - \mu_A(x) \tag{4.11}$$

<u>Example 4.8:</u>
$E = \{ x_1, x_2, x_3, x_4, x_5, x_6\}$; $A = \{ 0.13, 0.61, 0, 0, 1, 0.03\}$;
The complement set of A with respect to E is: $A' = \{0.13', 0.61', 0', 0', 1', 0.03'\}$ where the apostrophe signifies *Boolean* **NOT**. The membership vector of the complement of A is: $\mu_A'(x) = \{ 0.87, 0.39, 1, 1, 0, 0.97\}$

4.15 Fundamental properties of fuzzy sets

The fundamental properties of fuzzy sets lie in the following definitions of fuzzy multivalent logic. Assume that the fuzzy set A is a subset of universe of discourse E. Figure 4.8 shows that set A and its complement, A' are, in general, not *disjointed,* that is, their *intersection* is not the empty set. It is obvious that A and A' overlap, that they have a finite intersection $A \cap A'$:

$$A \cap A' \neq \square \qquad (4.12)$$

At the same time, the *union* of a fuzzy set and its complement are not equal to their universe of discourse. In this case, A and A' do not completely fill the universe of discourse E, there may something be left over, a so-called "underlap":

$$A \cup A' \neq E \qquad (4.13)$$

4.16 Some practical aspects

The foregoing sections covered various basic definitions as well as the terminology of both crisp and fuzzy theory. In particular, attention was given to various alternative methods of combining sets of both kinds. The question should arise as to the practical usefulness of these methods, that is, which one of these would apply to a particular control problem utilizing fuzzy logic. It is useful to give here a short preview of some of these concepts while deferring their detailed discussion to later chapters. A multi-input multi-output system can be characterized by a set of rules of the form

IF $var_1 = A$ <connective^t> $var_2 = B$ <connective^t> ... **THEN** $var_{01} = C$ <connective^t>...
<connective^s>
IF$var_1 = D$ <connective^t>$var_2 = E$ <connective^t>...**THEN**$var_{01} = E$ <connective^t>...
<connective^s>...

.....

.....

where A, B, C, D and E are crisp or fuzzy sets, and <connective^i> represents the particular fuzzy operator chosen to express the fuzzy inference or fuzzy implication desired. In fact, this <connective^t> operator combines sets A and B in the first rule and sets D and E in the second rule above. The combining of fuzzy sets within one rule is called *aggregation*. Similarly, <connective^s> combines the output sets of each rule in an operation called *composition*. The input variables appearing as conditional arguments under **IF** are referred to as *antecedents*, while the output variables under **THEN** are called *consequents* [1]. The rule set maps the input signals to the output the same way as the *Laplace* transfer functions do in linear control theory and the chosen way of combining inputs to produce an output, is equivalent to choosing a method of *system structure identification*. Thus the choice of the method of combining sets assigned to system variables has as important bearing on the plant/process controller structure. Due to the fact that many such fuzzy inference structures are available from the theory, various rule configurations must often be tried to optimize performance. Thus fuzzy controller design relies heavily on *empirical tuning* and this is no different than the tuning required for mathematical models of plants/processes of equal complexity. The difference is that fuzzy controllers need to be tuned like this only once during the design procedure and, in fact, such tuning is a integral part of the design procedure. There exists, however, a valuable body of experience and besides practical guidelines, certain rule configurations have crystallized as the best choice for specific kinds of control cases. One might say, that the design procedure for fuzzy controllers is also fuzzy, even though modern software-based development systems do provide a systematic design procedure resulting in excellent functionality and robust designs. These days, sophisticated interactive fuzzy software packages are used to perform the many iterations of tuning required.

One must not forget, that such rules may often result from an *interview* with an experienced human operator, in which case the designer's freedom of choosing the inference structure may be limited. On the other hand, in such cases less tuning is expected, in as much as the operator's experience already contains the best performance components embodied in the rules. The only tuning that remains to be done in such cases is that of the membership functions. The aggregation and combination of all fuzzy rules constitutes the so-called *knowledge base* of the controller which is the depository of all intelligence to carry out the desired control functions under the conditions specified. The following chapter will show how the knowledge base is generated in the form of a so-called *fuzzy relation*, analogous to the transfer function of a linear controller. Constructing the knowledge base is always the *first phase* in the design of an intelligent system. This is not unlike the determination of the transfer function (system identification) in linear theory. The *second phase* contains the actual use of the knowledge base by applying various inputs from the outside world as well as from the knowledge base to the system in order to produce an output. This is also similar to the use of the transfer function in linear theory.

4.17 The Fuzzy Approximation Theorem

The *fuzzy rule-based system*, $FRBS = (\mu_{ab}, R, T, S, DEF)$, is a family of fuzzy systems. It has membership functions μ_{ab} of positions a and widths b, a fuzzy rule base R, the *t*-norm, for fuzzy aggregation T (i.e. operations within one rule), the *s*-norm for fuzzy composition S (i.e. operations among rules), and the defuzzification method *DEF*. Defuzzification consists of the conversion of the fuzzy output into a single crisp output. (See Chapter 5).The main parameters of the *FRBS* are the number of fuzzy rules k and the positions a and widths b of the input and output membership functions. Of prime importance is that any given fuzzy system $FS \in FRBS$ is a *universal approximator* according to the following theorem:

Theorem: a fuzzy system is a universal approximator. Let the fuzzy rule-based system *FRBS* be the set of all fuzzy systems *FS* and $f: U \subset R^n \Rightarrow R$ be a continuous function defined on a universe of discourse U. For each $\varepsilon > 0$, there exists a $FS_\varepsilon \in FRBS$ such that

$$sup \{ / f(x) - FS_\varepsilon(x) /, x \in U \} \leq \varepsilon \qquad (4.14)$$

The proof is supplied, among others, by Castro [1993][2]. The operator *sup* signifies the *supremum*,[3] i.e.the least upper bound of the expression $/ f(x) - FS_e(x) /$ for all values of x in the universe U. Stating it loosely, Equation 4.14 expresses the idea that even the greatest difference between the function $f(x)$ to be approximated and any fuzzy set $FS_e(x)$ of the family *FRBS* approximating it, is always smaller than or equal to any arbitrarily chosen small positive number ε. Thus the function $f(x)$ can be approximated by a rule-based fuzzy system to any degree of accuracy that one wishes. From the theoretical viewpoint, a *FRBS* performs the same actions as any other approximation method. However, the fuzzy rules much more closely resemble the way humans explain general rules. Thus the fuzzy rule-based algorithm approximates the function $f(x)$ in a much more clearly understandable way and the fuzzy rule-based system *FRBS* provides a better model for the way humans think.

4.18 Fuzzy set-theoretical properties

Besides the techniques of combining fuzzy sets (and occasionally crisp sets with fuzzy sets[4]) covered in previous sections, one must also state the rules of performing set-theoretical operations in sequence. In addition, some properties of fuzzy sets, shared with crisp sets, are also to be discussed. Given universe of discourse E, and three fuzzy sets $A \subset E$, $B \subset E$, $C \subset E$, then the following applies:

Commutativity:

$$A \cap B = B \cap A$$
$$A \cup B = B \cup A$$

Associativity:

$$(A \cap B) \cap C = A \cap (B \cap C)$$
$$(A \cup B) \cup C = A \cup (B \cup C)$$

Idempotence:

$$A \cap A = A$$
$$A \cup A = A$$

Distributivity with respect to intersection:

$$A \cap (B \cup C) = (A \cap B) \cup (A \cap C)$$

Distributivity with respect to union:

$$A \cup (B \cap C) = (A \cup B) \cap (A \cup C)$$

Fuzzy set and its complement (*):

$$A \cap A' \neq 0$$
$$A \cup A' \neq E$$

Fuzzy set and the null set:

$$A \cap \square = 0$$

Fuzzy set and the universal set

$$A \cap E = A$$
$$A \cup E = E$$

Involution:

$$(A')' = A$$

De Morgan's theorem:

$$(A \cap B)' = A' \cup B'$$
$$(A \cup B)' = A' \cap B'$$

All the properties applicable to crisp sets are the same as the above, except the properties related to a fuzzy set and its complement marked above with (*)[5] .

4.19 Set operations in different universes of discourse.

The next task is to investigate *operations between sets belonging to different universes of discourse*. In the following, the term "space" will be used interchangeably with the term "universe of discourse". In control systems, mappings between input and output are our main concern. These mappings are between input variable sets $A(u) \in U$ and output variable sets $B(v) \in V$ through the conditional statement of inference:

$$A \Rightarrow B$$

or:

IF $A(u)$ **THEN** $B(v)$

which links the *antecedent* (condition) set A (defined *by* its membership vector $\mu_A(u)$, $u \in U$ with the *consequent* (result or action) set B (defined by its membership vector $\mu_B(v)$, $v \in V$).

4.20 Cartesian product of crisp sets

Let $A(u)$ and $B(v)$ be crisp sets, By taking u, one of the elements of A, and v, one of the elements of B to form an *ordered pair (u v)* one gets their *Cartesian product*,

$$P(u, v) = A(u) X B(v)$$

where the symbol "*X*" stands for the Cartesian product operator. Since both sets A and B are characterized by their respective membership vectors, their Cartesian product will be a matrix of crisp numbers:

$$P = A \ X \ B = \Sigma \ \mu_A(u) \ t \ \mu_B(v)$$

In practice, the *t*-norms *min* and *algebraic product* are mostly used . If $A1,...,An$ are crisp sets in spaces $U_1, ..., U_n$, then the Cartesian product of $A1,...,An$ is a crisp set in the product space $U_1 X U_2 X...X U_n$ with membership functions:

$$min\, [\mu_{A1\,X A2\,X ...X An}\,(u_1,u_2,...,u_n)] = min\, [\,\mu_{A1}(u_1),\, \mu_{A2}(u_2),...,\mu_{An}(u_n)\,),\text{ or:}$$

$$\textit{algebraic product:}\ \mu_{A1\,X A2\,X...X An}\,(u_1,\,u_2,...,\,u_n) = \mu_{A1}(u_1).\,\mu_{A2}(u_2).....\mu_{An}(u_n)$$

Example 4.9:

Let A and B be sets belonging to spaces U and V respectively.

$A = \{\, x_1,x_2,x_3\};\ \ A \in U$

$B = \{\, y_1,y_2,y_3\};\ \ B \in V$

The Cartesian product is a matrix:

$P = U X V = \{\, \{x_1,y_1),\, (x_1,y_2),\, (x_2,y_1),\, (x_2,y_2),\, (x_3,y_1),\, (x_3,y_2)\,\}$

Using the *t*-norm *min* the membership matrix of the Cartesian product is calculated numerically from:

$$\mu\,(x,\,y) = min\,[\,\mu(x),\,\mu(y)\,],\, x \in A,\, y \in B \tag{4.15}$$

Thus according to this definition the Cartesian product is an *intersection between sets belonging to different universes of discourse.*

4.21 Cartesian product of fuzzy sets

In this case, the membership grades, μ, are in $M = [0,1]$. If the spaces are, for example, $U_1 = \{x\}$ and $U_2 = \{y\}$, then the Cartesian product is $P\{x,\, y\} = U_1 X U_2$ with membership function $\mu_c(x,\, y)$ where each ordered pair is in $[0,1]$. In fuzzy set theory, the component sets of a Cartesian product are always universes of discourse, hence they are always crisp.

4.22 Relation

A Cartesian product $P = U\ X\ Y$ contains all ordered pairs as its elements. A set of ordered pairs, for example, may be a table of input-output values $u[k],\, y[k],\, k = 1,2,...,N$ of an *m*-input *p*-output discrete system. The *m* inputs $u_m[k]$ are in space U and the *p* output values $y_p[k]$ are in space Y, where k represents all possible values that occur in the respective spaces. If $k < N$, then such a subset is called a *relation:*

$$R \subset U\ X\ Y \tag{4.16}$$

This means that, in general, a relation contains all nonzero ordered pairs of the Cartesian product as well as some that are zero. In a relation, a rule exists that binds every output element $y \in Y$ with a corresponding input element $u \in U$. It is expected that this input-output table had been generated by some model of a physical process which determines a fixed relationship between input and output and thus satisfies the requirements of a relation calculated as

$$R\{ u_m[k], y_p[k] \} \subset U \times Y \quad (4.17)$$

4.23 Relation matrix

Let there be a relation R, defined from a set $X \subset U$ to a set $Y \subset V$ as $R \subset U \times V$, shown in Figure 4.9 in matrix form $U \Rightarrow V$. The entries in each matrix element, $\mu \in [0,1]$, are the membership values pertaining to each ordered pair x_i, y_j. Since the entries map the input universe to the output universe, the actual number in $[0,1]$ is also being referred to as the *mapping intensity* or strength of connection. The arrow shows the direction $U \Rightarrow V$ which is important because the relation $U \Rightarrow V$ is not the same as $V \Rightarrow U$. The relation $V \Rightarrow U$ is the *inverse* of relation $U \Rightarrow V$, obtained by interchanging rows and columns.

Example 4.10:
Let the matrix elements be: $\mu(x_1,y_1) = 0.3$; $\mu(x_1,y_2) = 0.7$; $\mu(x_2,y_1) = 1$; $\mu(x_2,y_2) = 0$; $\mu(x_3,y_1) = 0.5$; $\mu(x_3,y_2) = 0.2$. Figure 4.9 shows the relation matrix composed of sets X and Y, $X = \{x_1,x_2,x_3\} \subset U$, $Y = \{y_1,y_2,y_3\} \subset V$, $M = [0,1]$. This graph $G \subset U \times V$ is also called a *fuzzy graph.* If all membership grades were crisp, then it would be an ordinary graph. Figure 4.10 shows the fuzzy membership values displayed as various shades of gray.

4.24 Conditioned fuzzy sets

The fuzzy set $B(y) \subset E_2$ is said to be *conditioned* on a space E_1 if its membership function μ_B also depends on a parameter x, which is an element of space E_1: $x \in E_1$.

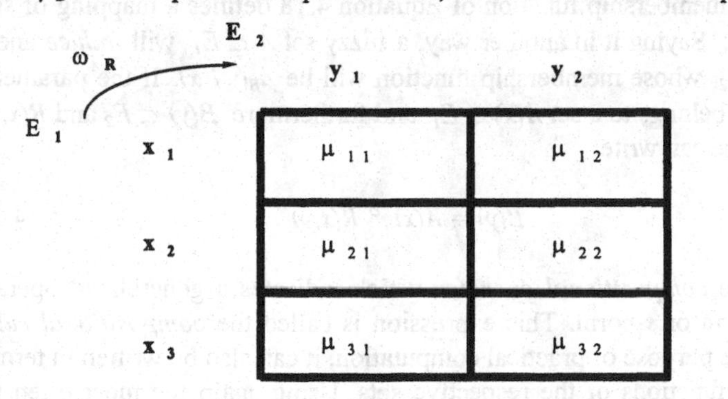

Figure 4.9. Relation matrix.

The conditional membership function is then

$$\mu_B(y \mid x) \quad (4.18)$$

where $x \in E_1$, $y \in E_2$. Equation 4.18 is thus the membership function of set B of elements y, given the parameter x. The notion of a *conditioned set* is an analog of the notion of a *function*: "If $x = a$ then $y = b$ by the function f, or $y = f(x)$.

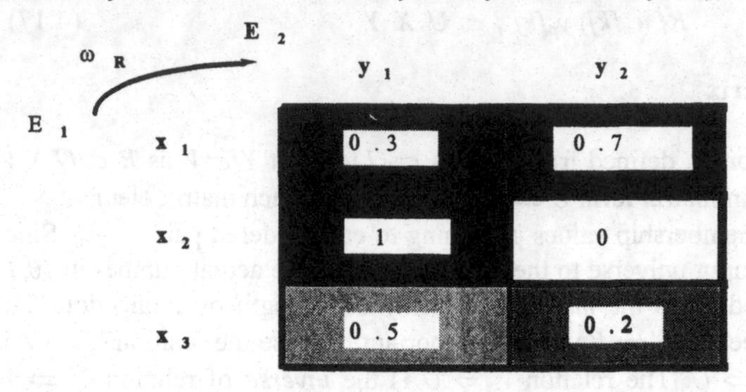

Figure 4.10. Relation matrix of Example 4.10

Similarly:

$$\textbf{IF } X = A \textbf{ THEN } Y = B \text{ through the relation } R\text{"} \qquad (4.19)$$

which can also be written as

$$A \underset{R}{\Rightarrow} B$$

4.25 The compositional rule of inference

The conditioned membership function of Equation 4.18 defines a mapping of space E_1 into space E_2.. Saying it in another way, a fuzzy set $A \subset E_1$ will *induce* another fuzzy set $B \subset E_2$ whose membership function will be $\mu_B(y / x)$. If the parameter x in Equation 4.18 belongs to a set $A(x) \subset E_1$, and furthermore $B(y) \subset E_2$ and $R(x,y) \subset E_1 X E_2$, then one can write:

$$B(y) = A(x) \circ R(x,y) \qquad (4.20)$$

where" ° " is the *compositional operator* which indicates a generalized operation similar to a *t*-norm or *s*-norm. This expression is called the *compositional rule of inference.* For the purpose of practical computation, it can also be written in terms of the membership functions of the respective sets. Using again the most often used compositional operators, the membership values of vector $B(y)$ may be calculated as follows:

Max-min composition: $$\mu_B(y) = \max_{x \in E_1} [\min [\mu_A(x), \mu_R(x,y)]] \qquad (4.21)$$

Max-product composition: $$\mu_B(y) = \max_{x \in E_1} [\mu_A(x) \, . \, \mu_R(x,y)] \quad (4.22)$$

In loose terms, the compositional rule of inference states that if the fuzzy relation representing a system, *R(x,y)*, is known, then the system's response (i.e. *B(y)*) can be calculated from a known excitation *A(x)*. There is an interesting analogy between this and the expression which represents the response of a linear system to a known excitation in terms of *Laplace* transforms:

$$Y(s) = U(s) \, . \, Z(s) \quad (4.23)$$

where *Y(s)*, the Laplace transform of the output, is an analog of *B(y)*; *U(s)*, the Laplace transform of the input, is an analog of *A(x)* and *Z(s)*, the *transfer function*, is an analog of the fuzzy relation *R(x,y)*. *Thus the fuzzy relation R(x,y) is both formally and functionally an analog of the transfer function of a linear system.* This is the basis of using fuzzy techniques for *modeling* complex systems and processes (i.e. systems identification) and for using such models to *compute their response to given excitations* (i.e. control).

Example 4.11:
Let there be two spaces $E_1 = \{x_1, x_2, x_3\}$ and $E_2 = \{y_1, y_2, y_3, y_4, y_5\}$. Define a fuzzy relation $R \subset E_1 \, X \, E_2$ as shown in Figure 4.10. Given a fuzzy set $A \subset E_1$ in terms of its membership vector: $A = \{0.3,\ 0.7,\ 1\}$ Determine the fuzzy set $B \subset E_2$ that is induced by the fuzzy set $A \subset E_1$ through the known fuzzy relation $R \subset E_1 \, X \, E_2$ shown in Figure 4.10. The *max-min* operation is carried out as follows: the *min* operation is similar to the rules of conventional matrix multiplication, where a row vector and a column vector are combined. The row vector's length must be compatible with the column vector's length. The resultant columnvector is calculated as shown in Figure 4.11.In turn, the *max* is taken:$max(0.3,0.7,0.2)=0.7 = \mu_B(y_1)$. The remaining fuzzy membership values of the vector *B* are calculated in the same manner, yielding: $\mu_B(y_2) = 0.3$; $\mu_B(y_3) = 0.7$; $\mu_B(y_4) = 0.4$; $\mu_B(y_5) = 1$;
Thus the fuzzy set (i.e. input) $A \subset E_1$ which induced the following fuzzy set (i.e. response) $B \subset E_2$ through the fuzzy relation $R \subset E_1 \, X \, E_2$ is: $B = \{\ 0.7,\ 0.3,\ 0.7,\ 0.4,\ 1\}$. Figure 4.11 shows the use of *max* in the compositional rule of inference to determine the fuzzy output vector from a fuzzy input vector and the fuzzy relation of Figure 4.10.

4.26 Fuzzy relation as a knowledge base

The foregoing has shown that a fuzzy relation behaves like a system model analogous to the *Laplace* transfer function which, if known, allows us to determine the system's response to a given excitation. In intelligent system terminology, the fuzzy relation is a *knowledge base*, that is, the depository of all intelligence related to a given system. Just like the transfer function of linear control theory, the fuzzy relation, if known, lets us to compute the system's response to given excitation

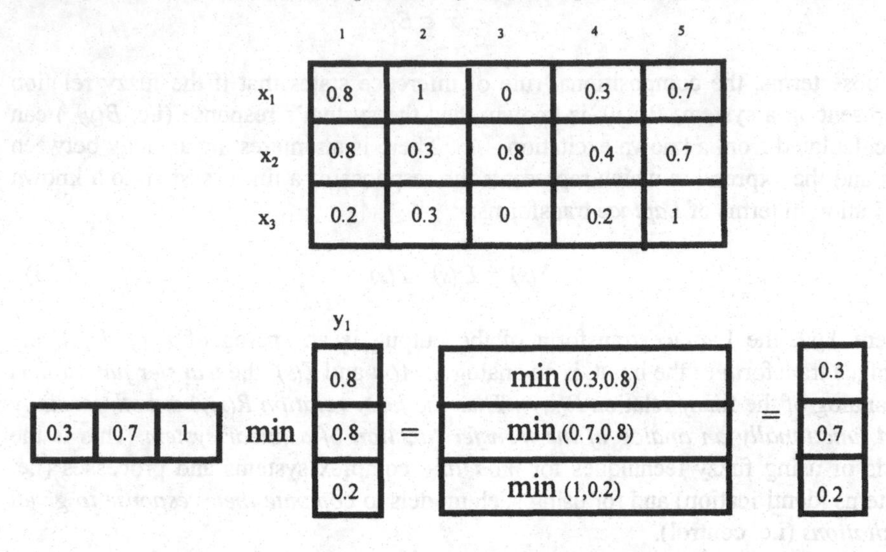

Figure 4.11. Using the compositional rule of inference

. The underlying difference between them is that unlike the transfer function, the fuzzy relation does not require linearity or time invariance as prerequisites for its use in computing the system's response. The control engineer is thus faces two tasks as follows:

(a) Determine the fuzzy relation. This is a *system identification* problem.
(b) Determine the response from the given excitation and the known fuzzy relation . This is a *control or estimation* problem.

4.27 Inverse composition

The *compositional rule of inference* (Equation 4.20) generates the system response from a given excitation through the known fuzzy relation. The *inverse composition problem* can be stated as follows: given a fuzzy relation $R \subset U X V$ where U and V are input and output spaces respectively. If given the system response (i.e. a fuzzy output set) $V_k \subset V$, find the excitation (i.e. a fuzzy input set) $U_k \subset U$ such that can generate the given response through the known fuzzy relation: $U_k \circ R = V_k$. A solution exists only if the so-called inverse composition $U_k = V_k \circ R^{-1}$ can be found.[6]. Fortunately, *a human operator performs inverse composition in an inherent manner* by applying the appropriate input to generate the desired output. As has been mentioned in Section 2.7, whenever fuzzy controllers are used to implement the control strategies of human operators the focus is on the human operator's behavior, that is, how he/she would adjust the control parameters for a given set of circumstances. In contrast to classical systems identification where the model of the plant or process under control is being identified, in fuzzy methodology *it is the operator whose model is being identified* while he/she is controlling the system, as shown in Figure 2.2. Thus in implementing a fuzzy controller that incorporates the intelligence of an experienced human control operator, the inverse composition problem is circumvented.

Notes

1 An awareness of these operators is important for those who wish to do specialized research in his subject. As far as industrial control systems are concerned, many of these operators are without much practical value.
2. However, other proofs also exist which are valid only for selected conditions of fuzzification and defuzzification methods, membership function types, etc. There exists no single universal approximation theorem which is valid for any arbitrary fuzzy system. Those that do exist can still support most engineering applications.
3. For those not familiar with this notation: the largest and the smallest of a finite set of real numbers $x_1, x_2, ..., x_n$ are called their maximum and their minimum respectively, denoted as $max[x_1, x_2, ..., x_n]$, $min[x_1, x_2, ..., x_n]$. If a sequnce of real numbers $x_1, x_2, ..., x_n$ is bounded from above, that is, if there are numbers v for which $v \geq x_n$, for every n, then there is among these numbers v a smallest one, say, v_1. It is variously called the least upper bound, the supremum, and sometimes the upper bounmd of the sequence $x_1, x_2, ..., x_n$: $v_1 = \sup[x_1, x_2, ..., x_n] = \sup x_n$. If, for example, the argument of *sup* is a sequence $x_1, x_2, ..., x_n = 0, ½, 2/3, ¾, ...$ then $\sup[0, ½, 2/3, ¾, ...] = \sup[(n-1)/n] = 1$ is the greatest upper bound of the sequence. If the expression has a maximum, then the supremum is equal to the maximum. Similarly, the greatest lower bound or infimum $u_1 = \inf[x_1, x_2, ..., x_n] = \inf x_n$ of a sequence bounded from below is defined exactly in the same way. The same terms are also applied to sets of numbers that are not given in the form of a sequence, for instance:

$$\sup_{a \leq x \leq b} f(x)$$

4. Note that in a *rule-based system* it is possible to mix crisp and fuzzy numbers as values of variables.
5. The above theorems have important practical applications in rule-based fuzzy systems, such as, for example, that within one rule the order in which fuzzy variables appear is of no consequence and similarly, the order of the rules themselves is not restricted to any specific order. A further analysis and proof of each statement is left to the reader.
6. However, the inverse composition of a fuzzy relation is difficult, lengthy and results in a non-unique solution. The mathematical solution of the inverse composition problem is mentioned here only for the sake of completeness.

References

CASTRO, JL: "Fuzzy Logic Controllers Are Universal Approximators". Dep. Comp. Sci. Univ. Grenada, Artif.Int.Tech.Rep.DECSAI-93101, June, 1993
DUBOIS D, PRADE H: *Fuzzy Sets and Systems*, Chapter 2; 191-192, Academic Press, New York, 1980.
KANDEL, A, LEE, SC: *Fuzzy Switching and Automata.* John Wiley and Sons, 1982.
KAUFMANN A.*Introduction to the Theory of Fuzzy Subsets, Vol. I.* Academic Press, 1975.
KOSKO, B: *Neural Networks and Fuzzy Systems.* Prentice Hall, 1992
LEE, CC: "Fuzzy Logic in Control Systems: Fuzzy Logic Controller, Parts I and II," IEEE Trans.Sys.,.Man.Cybern. 1990; 20; 2; 404-435.
PEDRYCZ, WS: *Fuzzy Control and Fuzzy Systems.* Report of Dept. of Math, Delft Univ.of Techn., Nederland,, 1982.
RAGADE , RK, GUPTA , MM: "Fuzzy Set Theory: An Introduction. In: *Fuzzy Automata and Decision Processes,* Eds. Gupta, MM, Saridis, GN, Gaines BR. North Holland, 1976.
TONG, RM: "A Control Engineering Review of Fuzzy Systems." Automatica, 1977; 13.

TONG, RM., BONISSONE, PP: A Linguistic Approach to Decisionmaking With Fuzzy Sets. IEEE Trans.Sys.,Man, Cybern. 1980; SMC-10;11.
ZIMMERMANN HJ: *Fuzzy Set Theory and Its Applications*. Second Ed., Kluwer Academioc Publishers, Boston ,Dordrecht,London,1991.
ZADEH LA: "Outline Of a New Approach To the Analysis of Complex Systems and Decision Processes." IEEE Trans. Sys,Man, Cybern.,1973; 3; 28-44

5 GENERIC STRUCTURE OF FUZZY CONTROLLERS

5.1 Basic configuration of a fuzzy controller

A fuzzy controller consists of four principal functional blocks, as follows:

- fuzzification interface,
- knowledge base,
- decisionmaking logic,
- defuzzification interface.

This controller structure represents a *transformation* from the real-world domain using real numbers to the fuzzy domain using fuzzy numbers, where a set of fuzzy inferences are used for decision-making, while the final stage provides an *inverse transformation* from the fuzzy domain into the real-world domain.

Fuzzification interface functions
The crisp (i.e. non-fuzzy) values of the input variables to this functional block usually come from sensors of physical quantities or from computer input devices. A scale mapping may be used to convert the actual input values into values that fit into preset universes of discourse for each input variable. In turn, the interface performs *fuzzification* whereby crisp input data are converted into fuzzy values in the interval *[0,1]* using membership functions that carry linguistic labels and are contained in the knowledge base.

Knowledge base
The knowledge base represents the model of the system to be controlled. It consists of *linguistic membership functions* and a *linguistic fuzzy structure* The *membership functions* expressed in linguistic terms are defined in numeric terms in a database, while the *rule structure* characterizes the control goals and control strategy of domain experts also expressed in linguistic terms. Thus the two principal components of a fuzzy system is its *structure* and the *fuzzy membership functions*.

Decisionmaking logic functions
The decision-making logic, embodied in the inference structure of the rule base, uses fuzzy implications to simulate human decision-making. It generates control actions inferred from a set of conditions.

Defuzzification interface functions
Defuzzification consists of deriving a single crisp value usable for a concrete real-world control action from inferred fuzzy output values. This single crisp value represents a compromise between the different fuzzy values contained in the controller output. This function is necessary only whenever the controller output is to be interpreted as a crisp control action, such as, for example, setting a switch into a discrete position, or moving a motor into a prescribed angular position. There exist systems, which do not require defuzzification because the fuzzy output is interpreted in a qualitative fuzzy way. See Note 2 , Chapter 3.

5.2 Fuzzy controller structures

In subsequent sections, the following kinds of *fuzzy controller structures*, each of which performs the functions above, will be discussed:

- rule-based fuzzy controllers,
- parametric fuzzy controllers,
- relational equation based fuzzy controllers.

The rule-based structure was the classical one to convert the linguistic statements of human operators into control rules. Its main shortcomings were as follows:

- in the case of difficult higher-order processes a systematic control strategy was not always easy to describe even by someone experienced with the process.
- there was no theoretical method to determine the minimum or even an adequate number of control rules.
- the controller structure is simple because the rule language may not be expressive enough to capture the heuristic knowledge needed to control complex processes.
- there are practical difficulties in correlating the verbal reports of operators with their mental activities controlling their behavior during an actual control task. That is, the extraction of rules and membership functions from operator interviews is a non-trivial task that might often need the services of experienced industrial psychologists.
- perhaps most importantly, there may not be enough experienced operators available whose heuristic knowledge could be utilized for fuzzy controller design.
- in summary, the above shortcomings gave an ad-hoc flavor to the whole process of developing rule-based fuzzy controllers. Initially, most fuzzy designs were made for simple home appliances where such ad-hoc techniques were not considered as hindrances. However, developing software design tools for more complex industrial systems to overcome these shortcomings became an unavoidable necessity. In Chapter 10 we will see how this was accomplished.

Nevertheless, the parametric and relational equation based structures structures were developed as an attempt to create a *systematic approach* to generate fuzzy rules from a given input-output data set. In other words, these approaches propose to lessen dependence on skilled human operators and rely on measurements instead. Of course, human experience and qualitative judgment would still be utilized when formulating membership functions in both the parametric and the relational equation based fuzzy structures, i.e. when constructing the afore-mentioned database which converts the linguistics expressions of human operators into numeric data in the interval *[0,1]*.

5.3 Fuzzy membership functions

Because of the commonality of membership sets used in the structures mentioned above, membership functions will be discussed before elaborating on different controller structures.

Fuzzy membership functions represent the fundamental aspects of all theoretical and practical realizations of fuzzy systems. A membership function database consists of a *graphical* or *tabulated* numeric function that assigns fuzzy membership values in the numeric interval *[0,1]* to the crisp values of a variable over its universe of discourse. (One should remember that the universe of discourse of a variable represents the numeric range of all possible values of real numbers that the specific variable can assume). In this book discussions are confined to discrete membership functions, even though illustrations show them being continuous, in order to relate directly to practical computer-based fuzzy controllers. Figure 5.1 shows some examples of membership functions erected over the universe of discourse of a crisp real variable, while the scale on the vertical axis represents the interval *[0,1]*.[1] Although the membership functions shown are *triangular* and *trapezoidal*, this is not a requirement. The number and shape of fuzzy membership functions are decided on the basis of design experience, the nature of the plant to be controlled, or an interview with an expert human operator who performs the control functions manually. In general, this is not a trivial task. However, some practical guidelines do exist as follows:

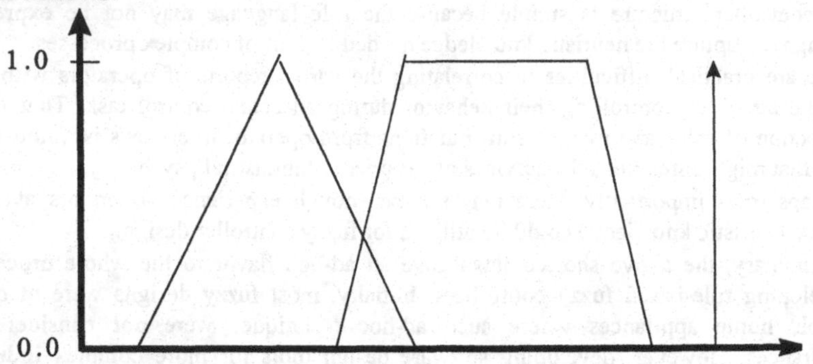

Figure 5.1. Examples of fuzzy membership functions

- A practical number of linguistic fuzzy sets (i.e. membership functions) is usually between *2* and *7*. The higher the number of sets the greater the accuracy but the computational overhead is also higher. For example, experience has shown, that for the Box-Jenkins data set widely used as a benchmark, going from *5* triangular sets to *7* enhances the accuracy by only about *15 %* and going above 7 does not yield any meaningful improvement in accuracy.
- The most frequently encountered shapes are *triangles and trapezoids*, in as much as they are easy to generate. In cases when a very smooth performance is of critical importance, *$\cos^2 x$*, *Gaussian*, *sigmoidal* and *cubic spline* (*S*-shape) functions can be used.
- Note that the membership functions do not have to be symmetrical or evenly spaced and that every variable may have a different set of membership functions with different shapes and distributions. For example, Figure 5.2 shows a typical set of membership functions for a position variable of a servo system. Note that near the central equilibrium point the membership functions are more dense, allowing a greater sensitivity for precise position adjustment, whereas farther away from the equilibrium point a coarser adjustment is acceptable. In such systems the universe of discourse runs from a negative to a positive extreme value, with zero at the equilibrium point.
- Another factor affecting accuracy is the amount of *overlap* between fuzzy membership functions. A minimum of *25%* and a maximum of *75%* have been established experimentally, with *50%* being a reasonable compromise, at least for the first trial of a closed-loop system. Of course, this overlap emerges as a result, not as an input parameter. The *inherent interpolation property* of fuzzy logic is partly due to the overlap between the fuzzy membership functions. Trapezoidal functions with a wide flat portion but steep overlapping sides may be used, like in Figure 5.3, where the fuzzy output is not sensitive to changes in crisp input values that lie right under the flat portions on the horizontal axis.
- Figure 5.4 shows a typical set of equidistant triangular membership functions. The usual linguistic labels in an industrial system for seven membership sets are:

NB = negative_big	*PS* = positive_small	*ZR* = zero
NM = negative_medium	*PM* = positive_medium	
NS = negative_small	*PB* = positive_big	

- However, other labels may also be used.

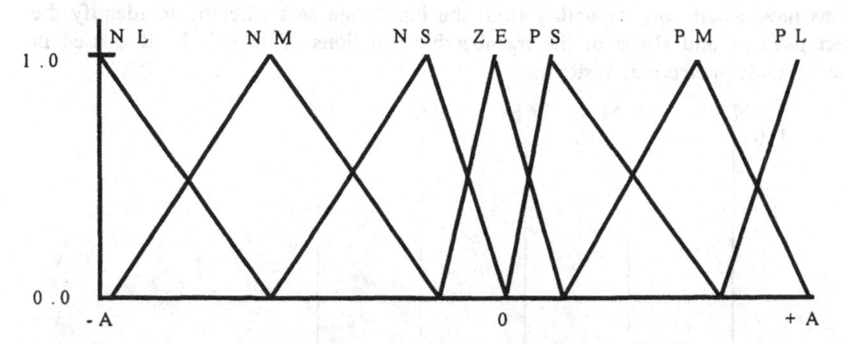

Figure 5.2 Membership functions for position variable of a servo system.

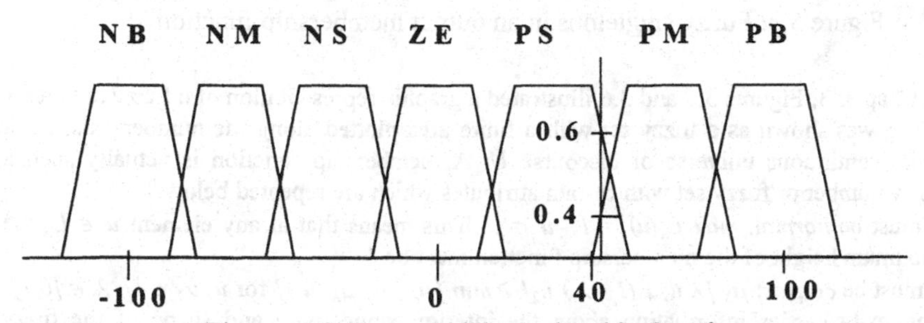

Figure 5.3. Trapezoidal membership functions

In the example below, there are only five membership functions with labels *L, ML, M, MH, H*:

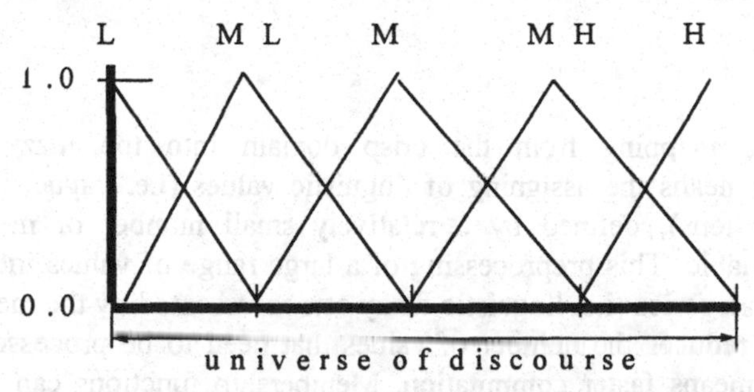

Figure 5.4 Equidistant triangular membership functions

In Chapter 3 we have seen that a single vertical spike on a horizontal number line can represent a crisp number, as shown in Figure 3.4 , with a height of, for example, unity. Such a spike can also represent a so-called *fuzzy singleton*, as in Figure 5.5 where it is used as a membership function for a fuzzy controller output. Fuzzy output singletons are used in conjunction with certain defuzzification methods (Figure 5.10) where they simplify the computations as will be seen in the detailed discussions on defuzzification.

- Due to the fact that setting up membership functions is a difficult design task, special techniques using *neural networks* are available to generate them automatically. Such systems have a *learning* capability from the input data sets whereby to identify the correct position and shape of the membership functions. This will be discussed in Chapter 9 under *neurofuzzy* systems.

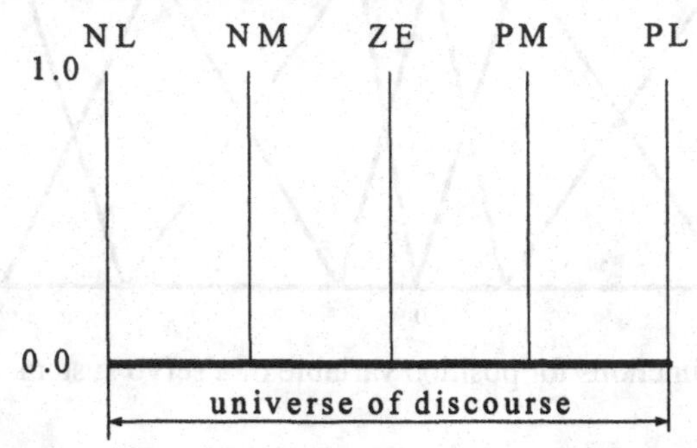

Figure 5.5. Fuzzy singletons in an output membership function

- In Chapter 3, Figures 3.5 and 3.6 illustrated a graphic representation of a fuzzy number X which was shown as a fuzzy set with a finite area plotted along the number axis, i.e. a finite continuous universe of discourse U. A membership function is actually such a fuzzy number or fuzzy set with certain attributes which are repeated below:
- X must be *normal*: $max\ \mu_X(u) = 1,\ \ u \in U$. This means that at any element $u \in U$, the maximum height of the membership function must be 1.
- X must be *convex*: $\mu_X\ [\lambda\ u_1 + (1 - \lambda)\ u_2] \geq min\ [\mu_X\ (u_1),\ \mu_X\ (u_2)]$ for $u_1, u_2 \in U,\ \lambda \in [0,1]$. Convexity carries information about the interior connectivity and shape of the fuzzy number. In other words, it prevents "holes" and "bays" within the boundary (Figure 3.7).
- Membership functions must be *complete*. This means that the set of membership functions assigned to a specific variable must cover the entire universe of discourse of that variable.It is also possible to use different membership functions for each variable, as shown in Figure 5.6.

5.4 Fuzzification

Fuzzification is a mapping from the crisp domain into the fuzzy domain. Fuzzification also means the assigning of linguistic values (i.e. vague, imprecise, qualitative descriptions), defined by a relatively small number of membership functions, to a variable. This preprocessing of a large range of values into a small number of fuzzy categories (i.e. linguistic fuzzy sets represented by the membership functions) greatly reduces the number of values that need to be processed. Fewer values processed means faster computation. Membership functions can also be a tabulated set of numeric values and a table look-up procedure, discussed in Section 5.6, can speed up fuzzification.

Example 5.1: Figure 5.3 shows a set of seven equidistant trapezoidal membership functions with linguistic labels whose universe of discourse has been normalized to *-100....+100*. Let us fuzzify the crisp value *+40*. A vertical line projected upwards from the point *+40* intersects the membership functions *PM* and *PS* respectively but none of the others. This should be interpreted as follows: the fuzzified equivalent of the crisp vale *+40* in this case belongs:

- to fuzzy membership function *PM* to the degree of *0.4*;
- to fuzzy membership function *PS* to the degree of *0.6*;
- to all others to the degree of *0.0;*

Thus the fuzzy vector equivalent to the crisp number *+40* is *{0,0,0,0,0.6,0.4,0}*. Note that the sum of the non-zero components is *1* because of the *50%* overlap.

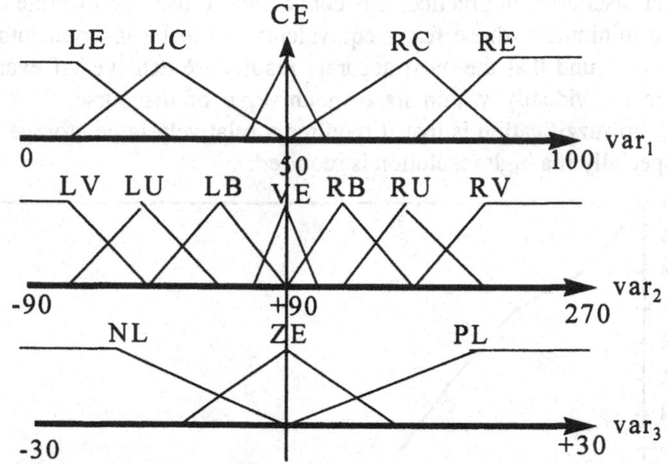

Figure 5.6. Different membership functions for each variable

5.5 Possibility vector

Example 5.1 showed the conversion of a single crisp value of a variable into a fuzzy set consisting of the same number of elements as the number of membership functions used in the fuzzification process and there was a *50%* overlap. If the number of membership functions is *M*, the fuzzification of an entire crisp vector $\{x_1,x_2,...,x_n\}$ would give rise to *M* fuzzy vectors $p_1, p_2,...,p_m$ called ***possibility vectors***, where each element would have two nonzero components (or just one in case the vertical projection of the crisp number to be fuzzified hits the peak of a membership function).

5.6 Table look-up fuzzification

In the foregoing it was assumed that the fuzzified equivalents of crisp values are calculated during run time, i.e. computing them from the equations representing the membership function. To shorten computation time, a table look-up fuzzification technique may be used as follows. The universe of discourse of each variable is usually normalized to a standard range adopted for the system on hand, thus the universe of discourse of each variable is known beforehand. This implies that the fuzzy values that correspond to the crisp values within this universe of discourse may be precalculated and inserted into a discrete look-up table. Thus instead of fuzzifying the discrete variables on line, the fuzzified values may be looked up in this table in real time, interpolating between the discrete values if necessary. This method speeds up fuzzification considerably. It is necessary to experimentally determine the lowest quantification of crisp data points that will still yield results of satisfactory accuracy. Figure 5.7 shows the fuzzification error corresponding to different numbers of table entries, assuming five equidistant triangular membership functions with a *50%* overlap. As can be seen the fuzzification error is less than *1 %* if the universe of discourse contains more than *150* crisp numbers. This result was obtained with 5 equidistant isosceles triangular membership functions covering their universe of discourse. In practice, it is convenient to use *256* discrete data points as a practical minimum, whose fuzzy equivalents are to be inserted into the look-up table. It was found that the most accurate results are achieved if every variable is normalized individually within its own universe of discourse. The drawback of table-look up fuzzification is that it requires a relatively large storage space for the tables, especially if a high resolution is required.

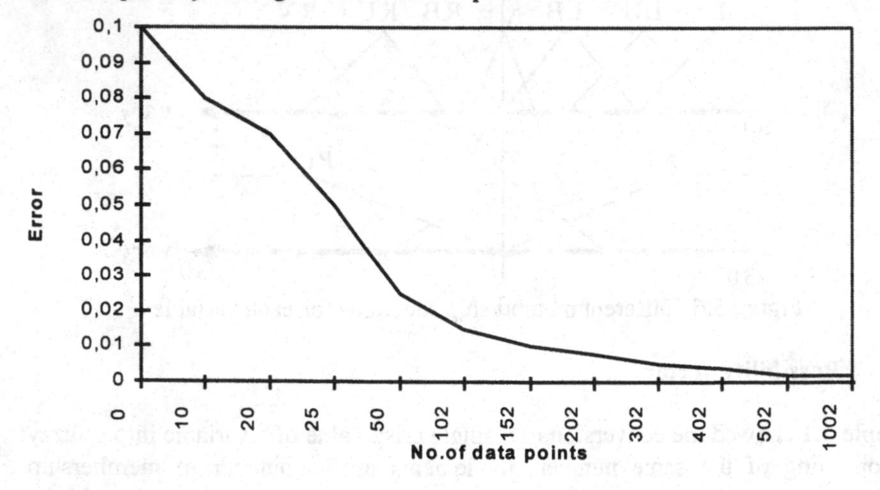

Figure 5.7. Table look-up fuzzification.

5.7 Defuzzification

In defuzzification, the value of the linguistic output variable inferred by the fuzzy rules is to be translated into a crisp value. The objective is to derive a single crisp numeric value that best represents the inferred fuzzy values (i.e. the possibility distribution) of the linguistic output variable. Defuzzification is an inverse transformation which retranslates the output from the fuzzy domain into the crisp domain.[2] To select the appropriate one from among several different defuzzification methods, one must understand the linguistic meaning that underlies the defuzzification process. The following defuzzification methods are of practical importance:

- Center-of-Area (*C-o-A*)
- Center-of-Maximum (*C-o-M*)
- Mean-of-Maximum (*M-o-M*)

5.7.1 Center-of-Area (*C-o-A*) defuzzification

The *Center-of-Area* method is often referred to as the *Center-of-Gravity* method because it computes the *centroid* of the composite area representing the output fuzzy term. Figure 5.8 shows the membership functions of a linguistic output variable.. Assume that there are five membership functions and that a specific fuzzy output (i.e. action or consequence) that emerged from the composite fuzzy inference rule is:

$$NB = 0.0; \quad NM = 0.0; \quad ZE = 0.2; \quad PM = 0.8; \quad PB = 0.0;$$

or, in possibility vector form: *{0.0, 0.0, 0.2, 0.8, 0.0}*

This fuzzy output is ambiguous, since two different actions, *ZE* and *PM*, have non-zero membership degrees. How can two conflicting actions, defined as fuzzy sets, be combined to a crisp real-valued output for driving a motor, for example?
One must remember that fuzzy sets are combined according to certain set-theoretical rules. Figure 5.8 shows the areas of *ZE* and *PM* combined by the *union* operator and thus their contour becomes the composite fuzzy output. In turn, the *C-o-A* defuzzification method computes the centroid of this area. The computation of the centroid proceeds as follows:

$$\sum_{j=1}^{p} y_j \; \mu(y_j) \; / \; \sum_{j=1}^{p} \mu(y_j) \qquad (5.1)$$

where $\mu(y_j)$ is the area of a membership function (such as, for example, *ZE* or *PM*) modified (i.e. clipped) by the fuzzy inference result (such as, for example, *0.2* or *0.8* respectively), and y_j are the positions of the centroids of the individual membership functions *ZE* or *PM* respectively. The numerator is actually the *momentum* of the particular individual membership function, with respect to the reference point on the horizontal scale. Equation (5.1) computes the composite centroid contributed by both of the membership functions indicated. In Figure5.8, $y_1 = -2.5$, $y_2 = 8$, $\mu(y_1) = 0.2$, $\mu(y_2) = 0.64$, thus $Defuz = [(-2.5)(0.2) + (8)(0.64)]/(0.2+0.64) = 4.62/0.84 = 5.5$
The *C-o-A* defuzzification method has two drawbacks. One of them is manifested only whenever the membership functions are not equal, as in Figure 5.9*a* Figure 5.9*b* shows that *LH* has a greater area than the membership function *ZE*, thus *LH* has a greater impact than *ZE* on *C-o-A* defuzzification. While the reason for having chosen a smaller area for *ZE* is that we wish to achieve a more sensitive control action near the central equilibrium point; yet this choice has a distorting influence on defuzzification. The second, more general drawback of *C-o-A* defuzzification is that it requires a high computational effort due to the numeric integration required.

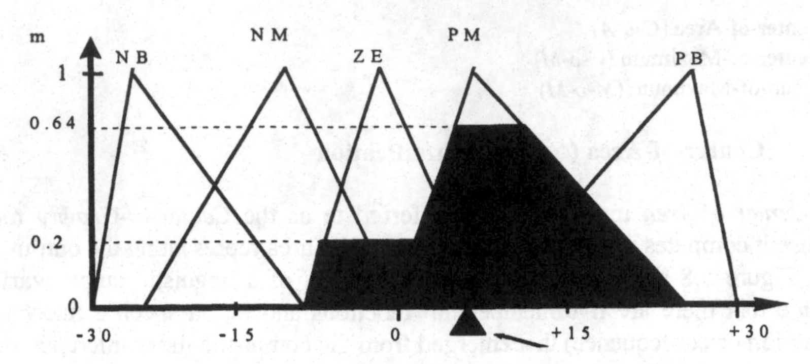

Figure 5.8. Center-of-Area (centroid) or *C-o-A* defuzzification.

5.7.2 Center-of-Maximum (*C-o-M*) defuzzification

In this method only the peaks of the membership functions (gray arrows) represented in the output variable's universe of discourse are used, the areas for the membership functions are ignored.. Refer to Figure 5.10 which displays membership functions for a linguistic fuzzy output variable.

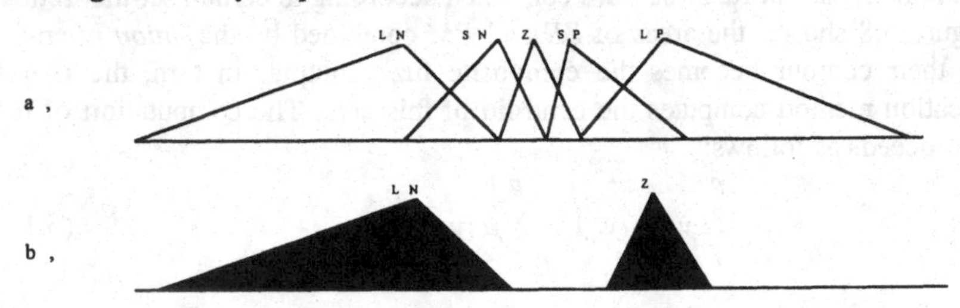

Figure 5.9. Bias in *C-o-A* defuzzification

The non-zero values of the output possibility vector are positioned on the corresponding peaks (black arrows). The defuzzified crisp compromise value is determined by finding the place of the fulcrum (representing the defuzzified value) where they are balanced. Thus the areas of the membership functions play no role and only the maxima (***singleton memberships***) are used. Assume that the fuzzy output vector is *{NB,NM,ZE,PM,PB}* = *{0.8, 0.2, 0, 0, 0}* and that the peaks of the membership functions occur at values *-0.25 (NB), -0.5(NM), 0 (ZE), 0.5 (PM), 2.25 (PB)* of the universe of discourse of the linguistic fuzzy variable. The computation proceeds as follows:

$$\sum_{j=1}^{p} y_j \ \mu(y_j) \ \ / \ \ \sum_{j=1}^{p} \mu(y_j) \qquad (5.2)$$

Defuz = *(0.8) (-2.25) + (0.2)(-0.5) + (0) (0) + (0) (0.5) + (0) (2.25) = - 1.9./ 1.0 =*
= -1.9

As can be seen, Equations 5.1 and 5.2 are virtually identical, except that Equation 5.1 uses the areas of each membership function, whereas Equation 5.2 uses only their maxima. As expected, the defuzzified results obtained will also be different. This approach yields the *best compromise* between the two possible outputs *NB* and *NM*.

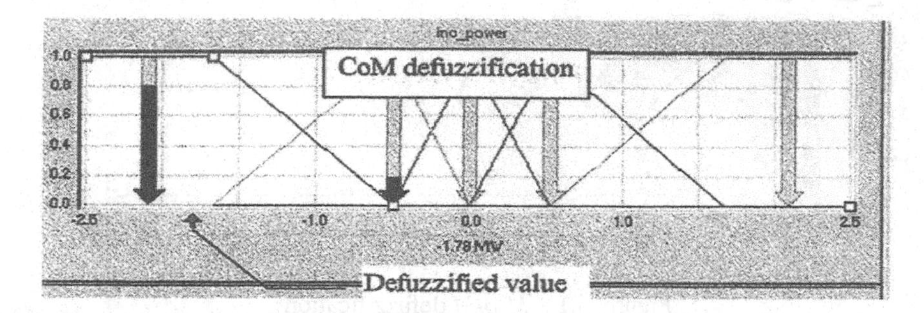

Figure 5.10. Center-of-Maximum (*C-o-M*) defuzzification

5.7.3 Mean-of-Maximum (*M-o-M*) defuzzification method

The *C-o-M* approach does not work whenever the maxima of the membership functions are not unique In this method, one takes the *mean of all maxima*:

$$max\ (\mu_I)\ /\ k \tag{5.3}$$

where $max\ (\mu_I)$ is the fuzzy output term with the highest membership degree and k is the integer number of such terms.If there is only one maximum than the *M-o-M* method (Figure 5.11) makes this one the defuzzified value in a "winner-take-all" situation. The *M-o-M* approach is also called the *most plausible solution.*

5.7.4 The concept of continuity in defuzzification

A defuzzification method is said to be continuous if an infinitesimally small change of an input variable can not cause an abrupt change in any output variable. The *C-o-M* and *C-o-A* defuzzification methods are continuous, while the *M-o-M* method is discontinuous because the *best compromise* can never jump to a different value for a small change in inputs while the *most plausible* solution is generally not unique and can thus jump over to a different value.

5.7.5 Which defuzzification method for which application?

In *closed-loop control applications*, the output of a fuzzy controller controls a process variable and jumps in the controller output can cause instabilities and oscillations, hence *C-o-M* defuzzification is recommended. However, in fuzzy *PI* controllers discussed in Chapter 6, an integrator placed between the controller and the process can assure that the control variable is kept continuous even if *M-o-M* defuzzification is used. In *pattern recognition applications M-o-M* defuzzification is the right choice, because sensor signal classification requires the most plausible result. The output possibility vector is the result of such a classification because it yields the similarity of the signal to the standard objects.

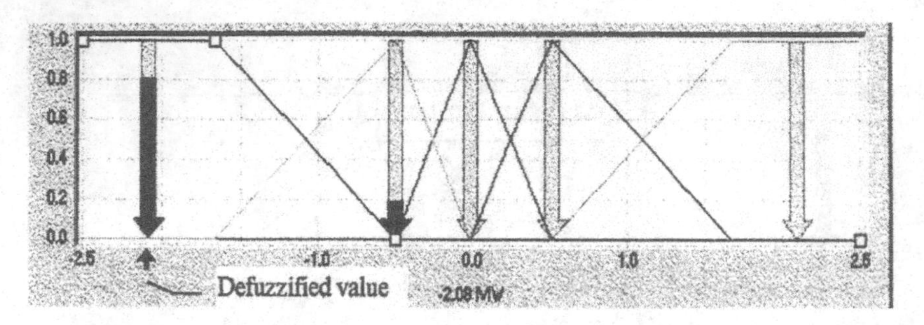

Figure 5.11. *M-o-M* defuzzification

Notes

1. The portion of a membership function that coincides with the universe of discourse plotted on the horizontal axis (i.e. where the membership is zero) is also very much a part of a membership function! In fact, membership functions could have been plotted on separate axes to show this to avoid misunderstanding. The reason why they were plotted on the same horizontal axis was to save space. (Refer to Figure 6.12).
2. The output of a fuzzifier and a defuzzifier connected in cascade will be only an approximation of the crisp input, thus fuzzification and defuzzification are not perfect inverse operations.

References

BAINBRIDGE L: "Verbal Reports As Evidence Of the Process Operator's Knowledge." In: Fuzzy Reasoning and its Applications, Mamdani EH, Gaines BR. Eds, Academic Press, 1979.

BOX GEP, JENKINS GM: *Time Series Analysis, Forecasting and Control.* Holden Day, San Francisco, 1970.

FuzzyTECH 3.1 *Explorer Manual and Reference Book*, Revision 310, Dec 1993.

HARRIS CJ, MOORE CG, BROWN M: *Intelligent Control: Aspects of Fuzzy Logic and Neural Nets.* World Scientific Publ., 1993, Singapore, ISBN 981-02-1042-6.

JANG JSR, SUN CT: "Neurofuzzy Modeling and Control". Proc.IEEE 1995;83;3;378-406.

KOSKO B: *Neural Networks and Fuzzy Systems.* Prentice Hall, 1992.

KOSKO B: *Fuzzy Thinking: the New Science of Fuzzy Logic.* Hyperion, New York, 1993.

LEE CC: "Fuzzy Logic in Control Systems: Fuzzy Logic Controller, Parts I and II," IEEE Trans. Sys.Man.Cybern. 1990, Vol 20, No 2, pp 404-435.

ROSS T:*Fuzzy Logic With Engineering Applications.*McGraw Hill, 1995.

TILLI TAW:"Practical Tools for Simulation and Optimization of Fuzzy Systems with Various Operators and Defuzzification Methods." EUFIT'93 Conference, 1993;256-262, Aachen, Germany.

VON ALTROCK C: *Fuzzy Logic and Neurofuzzy Applications Explained.* Prentice Hall, 1997.

WILLAEYS D, MALVACHE N: "Contribution to the Fuzzy Sets Theory of Man-Machine Systems. In: *Advances In Fuzzy Theory and Applications*, Gupta MM, Ragade RK, Yager R Eds. North Holland, 1979.

6 FUZZY CONTROLLERS

6.0 Introduction

This chapter describes the three alternative structures of fuzzy controllers entioned earlier in Section 5.2. The first one to be discussed is the rule-based fuzzy controller.

6.1 Inference rules in rule-based fuzzy controllers

In conventional logic there are two kinds of basic fuzzy implications, inference rules or associations: the *modus ponens* (affirmative mode) and the *modus tollens* (negative mode), both of which function on the basis of premises or conditions to generate a consequence.

	Premise 1:	$u = A$
modus ponens:	Premise 2 :	**IF** $u = A$ **THEN** $y = B$
	Consequence:	$y = B$

This is closely related to the *forward driven inference* mechanism used in expert systems.

modus tollens:

Premise 1: $y = not\text{-}B$
Premise 2: **IF** $u = A$ **THEN** $y = B$
Consequence: $u = not\text{-}A$

This is closely related to the *backward driven inference* mechanism used in expert systems. It will not be used in this book.

Expert systems that use forward driven (also referred to as forward-chained) inference usually use crisp variables or symbolic variables converted to crisp numbers.. As a result, they have to deal with a vast number of (i.e. several thousand) rules which is often several orders of magnitude greater than the number of rules of a fuzzy system (typically *20* to *100*) . In addition, in expert systems rules are activated in series, not in parallel. Their true purpose is to perform diagnostics of some kind, or act as an advisor. For these reasons, the computational overhead, structural complexity and memory demand of expert systems is rather high; they cannot run on microprocessor-based personal or industrial computers, and they are not amenable to real-time control. The following example will illustrate the use of an expert system.

Example 6.1:
Assume a relatively simple expert system used to diagnose your car and that you can enter an inquiry in English into a keyboard terminal, such as, for example: "CAR DOES NOT START". This symbolic input is coded as a crisp number via a table, say, *2000* and the antecedent portion of all rules compared with *2000* until a match is obtained. This particular rule with an antecedent coded as *2000* is said to be "*instantiated*". In turn, the consequent portion of this rule is used to again compare the antecedents of all rules until a new match is obtained, etc. This process is called *forward chaining*, i.e. the repeated use of *modus ponens*. The consequent portion of the last *instantiated* rule is then displayed on a terminal screen for the user as a reply to his query: "CHECK THE BATTERY".

A *fuzzy rule* or fuzzy inference relates fuzzy sets uses the so-called generalized *modus ponens* as follows:

Premise 1: "$u = A^{fuzzy}$ "
Premise 2: "**IF** $u = A$ **THEN** $y = B$"
Consequence: "$y = B^{fuzzy}$ "

where A and B are not necessarily the same as A^{fuzzy} and B^{fuzzy} respectively.

Example 6.2: Let there be the following rule:

IF temperature = high **THEN** boiler = highly stressed

where A^{fuzzy} = "high" and B^{fuzzy} = "highly stressed".

Assume Premise 1: A is less than *90* C° . Thus $A^{fuzzy} \neq A$ but A^{fuzzy} is similar to A.
Consequence: B = "dangerous". Thus $B^{fuzzy} \neq B$, but B^{fuzzy} is similar to B.

As regards the *modus ponens* of classical logic, a rule will be fired only if Premise 1 is exactly the same as the rule's antecedent and the result becomes of the rule's consequent.

As regards the *generalized modus ponens* of fuzzy logic, a rule is fired as long as there exists a *nonzero similarity* between Premise 1 and the rule's antecedent and the result becomes a consequent which bears a *nonzero similarity* of the rule's consequent.

Thus a fuzzy inference or rule is of the form:

$$\textbf{IF } X = A \textbf{ THEN } Y = B \tag{6.1}$$

where A and B are fuzzy sets and $A \subset X$, $B \subset Y$. A fuzzy controller contains many such fuzzy inferences, all of which are *activated in parallel*, i.e. at the same time. Thus a fuzzy controller "reasons" with *parallel associative inference*. When given an input, a fuzzy controller *fires* each rule in parallel but to a different degree, dependent upon a weight called *Degree of Support* which is a number in the interval *[0,1]* assigned to each rule to infer a conclusion or output. (The assignment of a fuzzy weight to a rule is referred to as a fuzzy associative memory or *FAM* system). The parallel operation gives fuzzy controllers their high processing speed. (For example, in a typical *3*-input *1*-output industrial fuzzy controller with *80* fuzzy rules, the cycle time to traverse the rule structure once may take less than a millisecond. Also refer to benchmarks in Chapter 10.) Fuzzy systems reason with linguistic sets instead of bivalent logical propositions. The general form of a fuzzy logical inference, fuzzy association, or fuzzy rule is:

IF <*conditions*> **THEN** <*conclusion*> (6.2)

or:

IF <*antecedent*> **THEN** <*consequent*> (6.3)

The <*conditions*> relate to linguistic fuzzy values of one or more variables: For example:

IF *pressure* = <very low>**AND** *temperature* <medium> **THEN** *valve* = <open a little>

Models of physical systems are *estimators* in that they estimate their output as a function of their input (see Chapter 2). A function $f: U \Rightarrow Y$ maps the input universe of discourse U to the output universe of discourse Y. That is, for every element $u \in U$ the function uniquely assigns an element $y \in Y$ and we denote this unique assignment as $y = f(u)$. Any physical, chemical or biological system estimates functions as they responds to stimuli and *associates* its responses with such stimuli.

In other words, it maps stimuli to responses or transforms inputs to outputs, and this transformation defines the *input-output function* $f: U \Rightarrow Y$ and indeed the system itself. Thus any system may be characterized by how it transforms input quantities into output quantities. Fuzzy systems can also estimate input-output functions, but by *heuristic* techniques:

Input variable $\Rightarrow$ *Fuzzy inference (rule)* $\Rightarrow$ *Output variable*

A human expert interviewed to help formulate the fuzzy rule set may articulate *linguistic input-output associations*. Thus fuzzy systems can produce estimates of a complex nonlinear system without resorting to mathematical models. The fuzzy method is a *mathematical model-free input-output estimation method.*

6.2 The choice of fuzzy inference

In Chapter 4, a multi-input multi-output system was characterized by a set of rules of the form

IF $var_1 = A$ <connective^t> $var_2 = B$ <connective^t>... **THEN** $var_{01} = C$ <connective^t>...
<connective^s>
IF $var_1 = D$ <connective^t> $var_2 = E$ <connective^t> ... **THEN** $var_{01} = F$ <connective^t>...
<connective^s>
......
......

where A, B, C, D, E and F are crisp or fuzzy sets, <connective^t> represents the general fuzzy intersection operator, and <connective^s> represents the general fuzzy union operator chosen to express the fuzzy inference desired. *The choice of the method of combining sets assigned to system variables has as important bearing on the fuzzy controller structure.*

There exists a valuable body of experience and certain rule configurations have crystallized as the best choice for specific kinds of control cases. Modern software-based development systems do provide a systematic design procedure using the appropriate choices resulting in excellent functionality and robust designs.

In *aggregation,* i.e. in combining fuzzy input sets within one rule, the *t*-norms *min* and *product* are most common, while in *combination,* i.e. in combining the fuzzy outputs of each rule, the *s*-norm *max* has been most practical (although some other connectives have also been used in rare instances). Thus one speaks of the *max-min* or *max-product* fuzzy controller structures. The *min* or fuzzy intersection operator implies an *AND* connective while the *product* is a fuzzy union operator implying an *OR* connective. The following section shows the detailed workings of a *max-min* type of fuzzy controller.

6.3 Fuzzy controller using *max-min* inference

Consider a simple *2*-input *1*-output fuzzy controller consisting of only two rules (Figure 6.1):

Rule 1: **IF** var_1 = *PS* **AND** var_2 = *ZE* **THEN** var_{out} = *NS*
Rule 2: **IF** var_1 = *ZE* **AND** var_2 = *ZE* **THEN** var_{out} = *ZE*

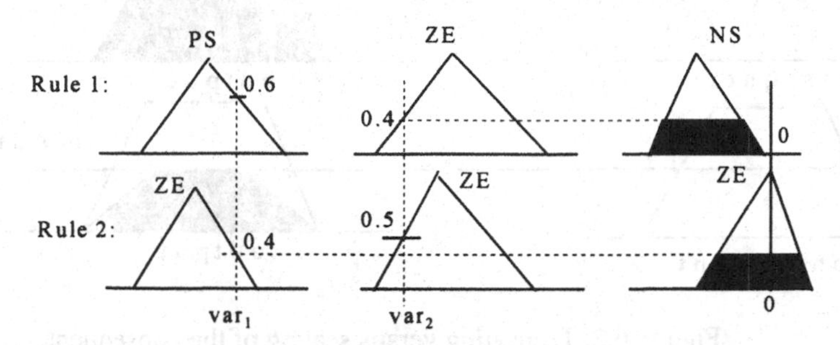

Figure 6.1. Aggregation in a two-rule system.

Step #1: **"AND"**
Input fuzzy numbers (i.e. membership sets) *PS* and *PM* are combined by using the *min* operator that corresponds to the **AND** connective according to the rules of fuzzy intersection. This operation is called *aggregation.* The two crisp inputs to the system are var_1 and var_2 plotted on the horizontal axis.

Consider Rule 1. A vertical line projected from var_1 to the antecedent fuzzy set *PS* cuts it at about *0.8*, while a vertical line projected from var_2 to the antecedent fuzzy set *ZE* cuts it at about *0.5*. Using **AND** we must take the min, i.e. the smaller value, and project a horizontal line to the right to the consequent fuzzy set *NS*, which is truncated at w_{NS} = *0.5*. This truncation is expressed as follows:

$$\mu_{NS'}(y) = min\ [w_{NS},\ \mu_{NS}(y)\ ] \tag{6.4}$$

The fuzzy set *NS'* is the truncated version of fuzzy set *NS* which represents the conclusion of Rule 1.

Rule 2 works the same way as Rule 1. The fuzzified value of var_1 is *0.4* while that of var_2 is *0.5*. Taking the smaller one again, the consequent set is cut at w_{ZE} = *0.4*:

$$\mu_{ZE'}(y) = min\ [\ w_{ZE},\ \mu_{ZE}(y)\ ] \tag{6.5}$$

The above operations are shown in Figure 6.1 and the consequent fuzzy set of Rule 2 is *ZE'*. It should be mentioned that if instead of *max-min* inference, *max-product* inference had been used, then the consequent fuzzy set would have been multiplied (i.e. scaled) by w_{NS} and w_{ZE} respectively.

Thus the *min* operation *truncates* the consequent fuzzy set while the *product* operation *scales* it. This is shown in Figure 6.2. It is obvious that the *product* operation provides an improved continuity and a smoother input-output function, in as much as the *min* operation creates nonlinear discontinuities in the fuzzy output.

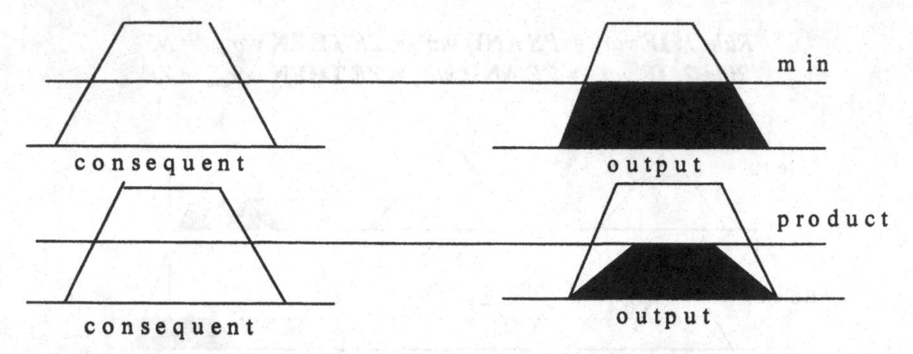

Figure 6.2. Truncating versus scaling of the consequent.

<u>Step #2:</u> **"OR"**

In this operation called *composition*, the fuzzy numbers corresponding to *NS'* and *ZE'* are combined by using the *max* operator that corresponds to the **AND** connective according to the rules of fuzzy union. The *max* operator creates the common contour or envelope of the two fuzzy sets. As has been mentioned in Chapter 5 , this operation converts the results of the fuzzy inference process into a single numeric value. In Figure 6.3, the composition operation as well as centroid defuzzification are used . It is interesting to note that the *min* operators produce no set interaction, since as soon as one input membership function used as shown above is not the smallest, it has no influence on the resultant output membership function, i.e. it "locks out" the others. This robustness is a decided advantage.

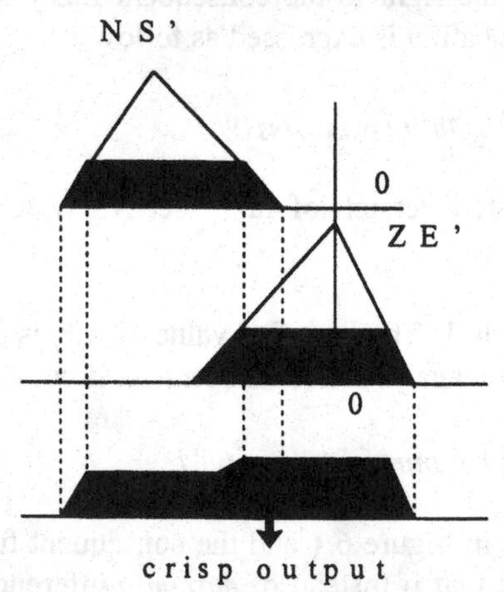

Figure 6.3. Composition and centroid defuzzification in two-rule system.

6.4 Fuzzy rule map

The fuzzy rule map is the same as the fuzzy relation introduced in Chapter 5. The fuzzy relation matrix entries were membership values in the interval *[0,1]*, whereas those of the fuzzy rule map are membership functions defined as, for example, *PB* (*positive_big*), *PM* (*positive_medium*), *NS* (*negative_small*), etc. As has been said before, the fuzzy rule map is the *knowledge base*, i.e. the depository of intelligence. The individual entries are filled in the rule matrix during fuzzy systems identification when typically a human operator is identified by means of an interview while he is controlling the plant or process. However, when such a rule map is implemented in a fuzzy controller, the situation is somewhat different. There could be intermittent disturbances, noise , or even a false input making a temporary entry into the rule map and this might trigger a false output. In a practical case, the empty cells are often filled with either alarm indications, or with the contents of the "nearest neighbor" cell as a default.

Example 6.2:
Let a fuzzy rule of a *2*-input *1*-output fuzzy system be:

$$\textbf{IF } v_1 = PM \textbf{ AND } v_2 = PS \textbf{ THEN } v_{out} = NS$$

Let the horizontal and vertical axes of the rule map represent v_1 and v_2 and the matrix entries v_{out} respectively. Then the fuzzy rule map is an array representing all combinations of input and output variables in terms of their linguistic fuzzy sets. In this example, output fuzzy set *NS* must be entered at the crossing of input fuzzy sets *PM* and *PS* as shown in Table 6.1. All other rules of the fuzzy rule structure must be entered similarly. Empty cells indicate that those particular input-output combinations were not used, or do not occur. This is similar to what human operators do when they select only the relevant rules. Fuzzy control faithfully reproduces human ability using simple non-redundant system descriptions.

Table 6.1. Fuzzy rule map for Example 6.2
(variable 1 = rows, variable 2 = columns)

	NM	NS	ZE	PS	PM
NM					
NS					
ZE					
PS					NS
PM					

6.5 Fuzzy interpolation

In the foregoing we have seen that the recommended maximum number of membership functions was *7* for input variables and thus the number of rows and

columns of an n-dimensional rule map hypercube would be 7^{n-1}. For example, in a 2-input 1-output case, the rule map has a maximum recommended dimension of $7 \times 7 = 49$ possible outputs which seems to be a rather coarse resolution, yet the control action provided is a continuous one. In this case, each entry in the rule map represents a rule with two antecedents and one consequent where the two antecedents describe a fuzzy region in the continuous state space of Figure 6.4 and there is a specific fuzzy consequent associated with each such fuzzy region.
Figure 6.4 depicts a 2-input 1-output state space, assuming that one of the input variables is plotted on the horizontal and the other one on the vertical axis respectively. Due to the overlap between membership functions which participate in defining individual fuzzy regions, the regions themselves overlap. When the fuzzy controller is exercised, each point in the thus quantized state space is affected by the actions of all the fuzzy consequents, thus describing a particular active fuzzy output region.The fuzzy composition combines the consequents and defuzzification defines an operating point on a hypersurface in the state space. The important point here is that an effectively quantized representation of the state space nevertheless yields a smooth action surface over the state space. The fuzzy controller provides an interpolation

- **from the finite number of fuzzy actions associated with the center points of the overlapping rectangular fuzzy regions (determined by the fuzzy input variables and their overlapping membership functions) by means of the inference connectives (i.e. *t*-norms) (Figure 6.4) , and**
- **from the defuzzification method used.**

Figure 6.4. Overlapping rectangular fuzzy regions

The design goal should be to achieve a smooth control surface with the minimum number of rules. Figure 6.2 implies that the *product* connective produces a smoother control surface than the *min* connective. The other design parameter that influences the smoothness is the membership function overlap which is obtained indirectly as a secondary result of membership function design rather than a parameter that can be set directly. The overlapping fuzzy regions that determine the smoothness of the control surface are also used in practice in conjunction with a *3*-dimensional diagram depicting the thus defined control surface itself (refer to Chapter 10). Note that the regions shown in white are not covered by any fuzzy rules of this specific rule base.

6.6 Composition which is not an *s*-norm

Kosko [1992] pointed out, that the *max* operation tends to produce a uniform distribution for the composition of fuzzy numbers as the number of combined fuzzy sets increases. A uniform distribution envelope has the same centroid, thus the greater the number of rules (each of which produces a contribution to the over-all area representing the main fuzzy output) the smaller the difference in the corresponding defuzzified value, because the centroid is almost the same. So unexpectedly, as the number of rules increases, the system sensitivity decreases. Instead of *max* composition (i.e. the *s*-norm or union via **OR**), *summing* was proposed whereby the fuzzy numbers representing the consequent sets of each rule would be added. However, the situation might not be so severe. In most multi-rule systems only a few rules of all possible ones make a contribution to the output at any given time.

Figure 6.5 illustrates *sum-min* or *sum-product* inference. For the sake of simplicity, it shows *1*-input *1*-output rules: for example **IF** A_1 **THEN** B_1. . The coefficients w_i are weighting numbers in the interval *[0,1]*, called the *Degrees of Support* that represent the relative importance of a particular rule in the rule set. The fuzzy numbers $B_1', B_2', \ldots, B_m'$ comprise the truncated (*sum-min*) or scaled (*sum-product*) membership sets which are summed into a single unified area whose centroid is to be calculated as the defuzzified value when using *C-o-A* for defuzzification. The summing method diminishes in importance whenever *C-o-M* or *M-o-M* defuzzification is used because in these approaches the areas of the consequent sets are ignored. Nevertheless it is available as an option in most up-to-date fuzzy controller software development systems.

6.7 Advantages of rule-based fuzzy industrial controllers

Rule-based fuzzy controllers have a number of practical advantages which made them the most popular configuration used by fuzzy controller software development systems.

- First and foremost, they can be implemented by a rule-based system, using *expert knowledge of human operators* formulated in linguistic terms. This is the basic reason why fuzzy controllers have been adopted in industrial control systems.
- Parallel operation is fast. The fuzzy controller completes the processing task without involved calculations, thus processing speed is improved. Ultra-high processing can be attained, if necessary, (although rarely required in an industrial application) by using hardware-based controllers such as *transputers*, that realize the full potential of parallel processing. In addition, dedicated fuzzy microprocessor chips have been used in high-

volume mass-produced equipment or special high-speed applications. However, in most industrial applications, software-based fuzzy controllers are definitely fast enough to handle the tasks on hand, yet remain flexible and transparent to the designer.

- Transparency refers to an important attribute of modern fuzzy software development systems, whereby the designer does not do any programming; he/she only manipulates graphic objects on the computer screen and when finished, the control program is written automatically in the language required by the control platform used. This will be discussed in Chapter 10 in more detail. Fuzzy control rules are easy to understand by maintenance personnel, in as much as they are patterned on the basis of human common sense and the effect or outcome of each rule can be easily interpreted. Hence training costs for maintenance personnel are reduced.
- All of the control functions associated with a rule can be tested individually. This improves maintainability because the simplicity of the rules allows the use of less skilled personnel.
- Individual rules combine to form a structure whereby complex control can be executed in a strict way. This cooperation of rules provided by parallel processing allows fuzzy logic to control complex systems using simple expressions. In addition, rules can be added for alarm conditions which fire rarely and add very minimal processing time to the cycle time, yet have a decisive influence when the time comes.

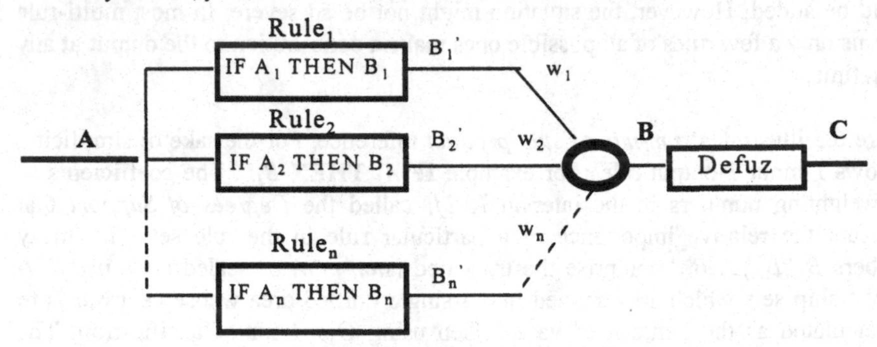

Figure 6.5. Fuzzy *sum-min* inference

- Many reports on systems implemented in practice show that fuzzy controllers are inherently *robust*, i.e. resistant to external disturbances and disturbances due to internal component wear and aging. Conventional systems process complex equations in sequence and if an error were to occur in even one of them, the final result would be totally unreliable. In a fuzzy controller each rule is processed independently, thus its effect on the final result is minimal. This means that a partial system failure may not significantly degrade the controller's performance.
- Experience shows that in contrast to conventional controllers, the development of fuzzy controller prototypes may take as little as a few hours or days. This may turn out to be a decisive advantage when the manufacturer wants to bring the fuzzy product to the market within the shortest possible time. Of course, whether or not this promise can be fulfilled depends on the complexity of the product.
- Experience shows that fuzzy controllers are often cheaper to produce, are more reliable than their conventional counterparts, and are insensitive to even large parameter variations. For example, in induction motor speed control, the rotor resistance can change by as much as *70%*, yet the fuzzy speed controller would still remain within the rated design speed tolerance (refer to a specific application described in Chapter 11). In

electric power applications, the cost and robustness factor becomes the decisive factor (rather than the use of expert knowledge) that militates in favor of fuzzy controllers.

6.8 *PID* controllers

A feedback controller generates an output which is detected by a sensor and used to apply a corrective action to the process or plant so as to drive a measurable process variable towards its desired value known as the *setpoint*. The difference between the desired value and the actual value of the measured process variable , i.e. the *error* drives the controller. An error can occur as a result of a *changed setpoint* or a *changed load.* Feedback controllers can be electrical, mechanical, hydraulic or pneumatic. The majority of industrial feedback controllers in use are of the electrical/electronic *Proportional-Integral-Differential (PID)* type, implemented mostly by programming a *Programmable Logic Controller (PLC)*. The flexibility, low cost and robustness of *PLC*s and the availability of functional hardware blocks such as, for example, central processing units *(CPU)*, counters, timers, arithmetic units, scalers, comparators, analog and digital inputs, stepping motor and relay drivers, communications modules, parallel and serial interface modules, integrators, differentiators, etc.) make it possible to program a *PLC* in many different ways. The program is stored in the program memory in the form of a *list* of individual statements, a control system *flowchart* which looks like a logic block diagram, or a *ladder diagram* which looks like a relay contact network. New *ISO 9000* standards make it possible, and indeed require, to program *PLCs* in common computer languages such as *C++™*, *Visual Basic™*, etc. developed on off-line *PCs*. The ubiquitous presence of computer technology prescribes the use of discrete digital *PID* controllers, and a *PID* controller becomes just another program in computer memory. The continuous error signal at the controller's input is sampled and converted into digital signals, while the discrete controller's digital output is re-converted into a continuous analog signal fed into the plant/process.

The *PID* controller is a *time-domain* controller whose continuous input-output function for position control is:

$$c(t) = K_c\, e(t) + (K_c/\tau_I) \int_0^t e(t)\, dt + K_c\, \tau_D\, de/dt \tag{6.6}$$

and its discrete form:

$$c_n = K_c\, [\, e_n + (T/\tau_I) \sum_{k=0}^{t} e_k + (\tau_D/T)\, (e_n - e_{n-1}) \tag{6.7}$$

Figure 6.6 shows the block diagram of a general discrete digital system. The analog signal $e(t)$ is converted by the *A/D* (analog-to-digital converter) to a discrete sequence of numbers, $m[kT]$, $k = 1,2,\ldots,N$. This requires that the analog signal be sampled at intervals $t = kT$. A sample-and-hold circuit holds the last sample value constant for the *A/D* converter until the next sampling interval. The discrete sequence $m[kT]$ is then processed by the control algorithm to obtain the control sequence $u[kT]$. In turn, this sequence is converted by a *D/A* (digital-to-analog) converter to the analog signal *u(t)* which drives the process. The *PID* control algorithm is just a software item that runs in the computer shown.

6.9 Problems with plant/process modeling

PID controllers are automatic controllers which work well if the plant/process they control is reasonably linear, i.e. a change in process input generates a proportional change in process output. There exist a number of *PID* tuning methods by *Ziegler-Nichols, Cohen-Coon* and others. *PI* (i.e.two-term) controllers are used more frequently than *PID* controllers and the remarks below are equally true for them.

The *PID* (i.e.*three-term)* controller determines the current value of the error, the cumulative error over a recent time interval (integral or past history of the error), and the currently predicted value (the derivative of the error, i.e. the tangent of the error's time function). Each of these terms, i.e. the proportional, the integral and the derivative term, is multiplied by a weighing constant which makes it possible to individually adjust them during a tuning process. Determining the best mutual adjustment of these weighing constants depends on the process/plant behavior required, such as, for example, fast rise time, minimal overshoot, zero steady-state error (i.e. high accuracy in approximating the setpoint). The proportional part improves the sensitivity to parameter variations, the integral part improves the steady-state accuracy and the derivative part improves the stability of the system by increasing the damping. Even a long (but constant) dead time can be compensated for by tuning the proportional and the integral weighing constants to a minimum step function error.

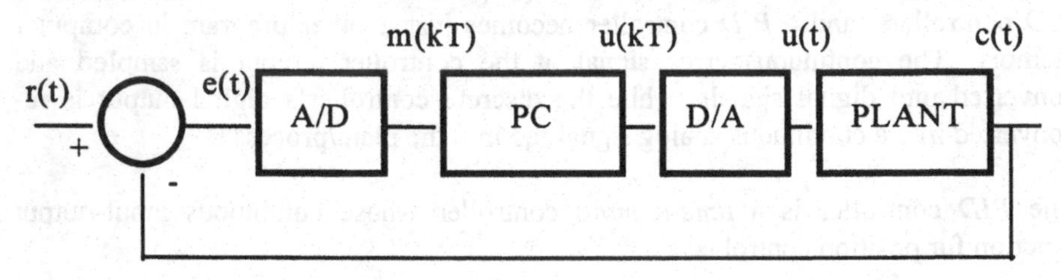

Figure 6.6. Block diagram of general discrete control system..

The *PID* controller is based on the assumption that the error is a linear function of time. If the process input-output relationship is mildly nonlinear, periodic tuning of the controller parameters is required. Mildly nonlinear means that the process operating point moves slightly along the input-output function which, within a narrow region, can be considered linear. However, in the case of substantially nonlinear processes, or whenever highly nonlinear control elements or actuators (such as, for example, control valves) are used in the feedback loop, or when mathematical process modeling (i.e. establishing the input-output function of the process) runs into difficulties because of complexity or poor knowledge of the process, automatic *PID* controllers give poor performance. In such cases, the only recourse is to keep using experienced human control operators that can make manual adjustments. Difficulties encountered in the mathematical modeling of a process may be classified as follows:

- Poorly understood chemical or physical processes

For accurate modeling, certain information about the process is required which is not only hard to come by, but at times may lead to contradicting requirements.

- Imprecisely known parameters

The parameters as well as their changes in time are needed.

- Dead time

Is another critical parameter which is imprecisely known and changes in time. Poor knowledge of dead time and its expected variations can lead to serious closed-loop stability problems.

- Size and complexity of model

Mathematical models of complex systems are cumbersome and difficult to use. Beyond a certain degree of complexity models lose their value and become unattractive.

- Nonlinearities

A linear system obeys the superposition principle, whereby if $c_1(t)$ is the system response to $r_1(t)$, and $c_2(t)$ is the response to $r_2(t)$, then the system's response to $a_1r_1(t) + a_2r_2(t)$ is $a_1c_1(t)+ a_2c_2(t)$. However, the superposition principle does not apply to nonlinear systems and mathematical operations used for analyzing linear systems are not applicable to nonlinear systems. Nearly linear systems where the deviation from linearity is not too large and the nonlinear input-output function is smooth and free of sudden jumps, within a small region of a variable around the operating point can be considered linear. Nonlinear processes, by definition, have an input amplitude-dependent gain. In a closed loop if the gain increases with the input amplitude, disturbances may induce instability. Nonlinearities that increase the loop gain at low amplitudes tend to cause continuous nonsinusoidal oscillations (limit cycles) of fixed amplitude and period. Limit cycles may be stable or unstable, and may be driven to oscillate with either very small or very large signals only. Figure 6.7 shows the input-output function of a *saturation* curve which clearly shows then output's dependence on the input amplitude. It also illustrates the input-output function of a saturation curve with a *dead zone, as well as backlash* or mechanical hysteresis which is due to the difference in motion between an increasing and decreasing output. The source of backlash is the looseness inherent in mechanical gearing. Electromagnetic hysteresis is similar.

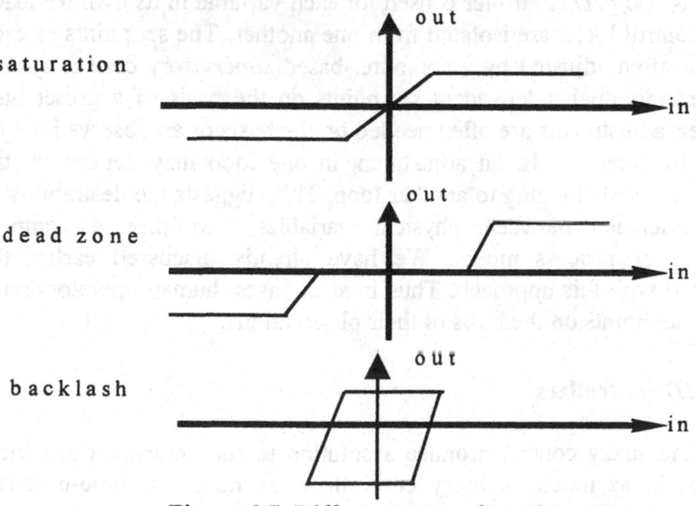

Figure 6.7. Different types of nonlinearities

How does a *PID* controller-driven nonlinearity manifest itself? The *PID* controller needs to be tuned for optimum response, that is, fast rise time, minimal overshoot, zero steady-state error, by means of its proportional gain K_c for the *P*-component, integration time constant τ_I for the *I*-component, and differentiation time constant τ_D for the *D*-component. When the plant/process is linear, these adjustments are independent from one another and can be tuned one-at-a-time. In the nonlinear case, however, they affect one another, and the controller becomes difficult or even impossible to tune properly. In industrial control, the controller drives a final control element such as, for example, a control valve whose input-output function (i.e. the drive from the controller versus the output flow rate) may be nonlinear. In fact, control valves are available with a number of optional input-output characteristics (determined by the geometric shape of the valve plug's surface) and in certain cases the valve characteristics are chosen deliberately by the system designer to be nonlinear. In most valves the operating power is usually pneumatic or hydraulic, but the control medium is electric or electronic. For example, pneumatic valves are among the most commonly used final control elements and it can be shown that under certain conditions they exhibit second-order input-output functions. Thus even if the controller and the process under control were linear, the final control element that drives the plant/process, might introduce a nonlinearity in the control loop.

- Harmonic distortion

A linear system with a sinusoidal input produces a sinusoidal output of the same frequency, although the amplitude and phase may differ from those of the input. However, a nonlinear system output usually contains additional frequency components, i.e. the system generates harmonic frequencies that were not present in the original input.

- Stability

The stability of a linear system depends only on the initial conditions and system parameters. However, the stability of a nonlinear system depends on the initial conditions, the system parameters, and the nature of the input signal. For example, a nonlinear system may be stable for one kind of input signal and unstable for another kind.

6.10 Use in multivariable control systems

PID controllers are single-input single output controllers. In industrial multivariable control applications one *PID* controller is used for each variable in its own feedback loop, that is, the control loops are isolated from one another. The set points of each *PID* controller are often adjusted by a computer-based *supervisory control system* which adjusts the individual independent set points on the basis of a preset time schedule. However, adjustments are often needed on the basis of an observed output rather than time. In other words, an adjustment in one loop may depend on the observation of a variable belonging to another loop. This suggests the desirability to exploit interdependencies between physical variables, requiring a complex mathematical plant or process model. We have already discussed earlier the problems associated with this approach. Thus in such cases, human operators must be used adjust the set points on the basis of their observations.

6.11 Fuzzy *PID* controllers

Fuzzy modeling and fuzzy control promise a solution to the control of nonlinear industrial systems, in as much as fuzzy controllers are nonlinear time-invariant dynamic-free feedback controllers.

The rule format of fuzzy controllers discussed earlier was:

$$\textbf{IF } v_1 = XX \textbf{ AND } v_2 = YY \textbf{ THEN } v_{out} = ZZ \tag{6.8}$$

The rule format of the fuzzy *PID* controller is:

$$\textbf{IF } v = XX \textbf{ AND } dv = YY \textbf{ THEN } v_{out} = ZZ \tag{6.9}$$

or, in other words, if one input is *XX*, then the other one is *YY*, the *rate of change* of *XX*.

6.11.1 Fuzzy proportional controller

Consider a *1*-input *1*-output discrete time proportional controller:

$$u = K_p\, e \tag{6.10}$$

where *e* is the error between a set point (i.e. reference) signal and the plant output. The fuzzy control rule is:

$$\textbf{IF } error = E_i \textbf{ THEN } control = U_i \tag{6.11}$$

where E_i and U_i are linguistic membership functions assigned to variables *e* and *u*. In this case, the fuzzy controller has a single input *e*.

6.11.2 Fuzzy proportional-integral (*PI*) controller

Let *de* be the change of the error *e* and *du* be the change of the output *u*. In the time domain:

$$du = K_p\, e + K_I\, de \tag{6.12}$$

The definition of the fuzzy rule is:

$$\textbf{IF } error = E_i \textbf{ AND } change\ of\ error = dE_i \quad \textbf{THEN } change\ of\ control = dU_i \tag{6.13}$$

This fuzzy controller has two inputs: the error and its first derivative. Note that the change of control expression needs to be integrated before it can be used to drive the plant/process.

6.11.3 Fuzzy proportional-integral-differential (*PID*) controller

The time domain expression of a *PID* controller is:

$$du = K_p\, e + K_I\, de + K_d\, d^2 e \tag{14}$$

where d^2e is the *change of the change of error*. This suggests using a fuzzy controller with three inputs e, de, d^2e, and one output dU governed by rules of the form:

IF *error* = E_i **AND** *change of error* = dE_i **AND** *change of change of error* = d^2E_i
THEN *change of control* = dU_i (6.15)

Note that here again the change of control expression needs to be integrated before it can be used to drive the plant/process. In addition, the weighing constants can also be used in conjunction with fuzzy *PID* controllers which makes them similar to the conventional ones in terms of tuning and dead-time compensation.

6.11.4 General remarks about fuzzy *PID* controllers

The two-input fuzzy *PI* configuration (Figure 6.8) has proven to be the most popular one and while it has given good results in many applications, its performance depends, to a very great degree, on its tuning, which must be accomplished in a trial-and-error manner. In fact, in designing any fuzzy controller many more choices and options exist than in the case of conventional controllers.

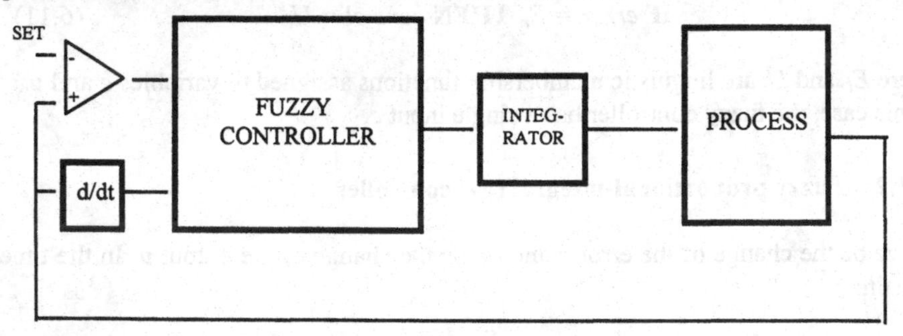

Figure 6.8. Fuzzy *PI* controller

The design and optimization (i.e. *tuning*) of a fuzzy system is burdened by the

$$k \times k_1 \times r \times r_1 \times r_2 \times m \times p \times d \qquad (6.16)$$

degrees of freedom, where m = number of input variables; p = number of output variables; k = number of membership functions for each variable; k_1 = shape of membership functions for each variable; r = number of fuzzy rules; r_1 = choices of inference expressed in the fuzzy rule structure; r_2 = degree of support associated with each rule; d = choice of defuzzification method. Many of these choices are based on existing empirical data and design guidelines and existing fuzzy controller software development packages make it practical to try out many options in an efficient and rapid manner either on appropriately simulated plants or *in situ* on the controlled plant itself.

The charge is often made against fuzzy control systems that their design is *ad hoc* and no systematic design procedure exists. Proponents of these views tend to forget that no design procedure, however systematic, is valid unless the designer is familiar with the physical or chemical plant or process to be controlled, including its problems and vagaries. *No systematic design procedure applied mechanically to a real-life system can ever generate a satisfactory control system in the absence of any previously known information.* The more the designer knows about his system (even if only qualitatively) the easier he will find to choose the parameters correctly, even if in conjunction with a fuzzy controller software development system that enables him to evaluate many equivalent solutions within a short time.

6.12 Fuzzy supervisory controller

The problem of using *PID* controllers in a supervisory control system was outlined earlier , where it was observed that single process variables are controlled by *PID* controllers while supervisory control is done by human operators. Fuzzy logic can provide an efficient solution to this problem whereby fuzzy supervisory multivariable controllers are designed on the basis of operator experience rather than mathematical models and the fuzzy supervisory controller controls the set points of the *PID* controllers. Figure 6.9 shows that each single process variable is kept constant by a *PID* controller while the set points for the *PID*s derive from the fuzzy supervisory controller.

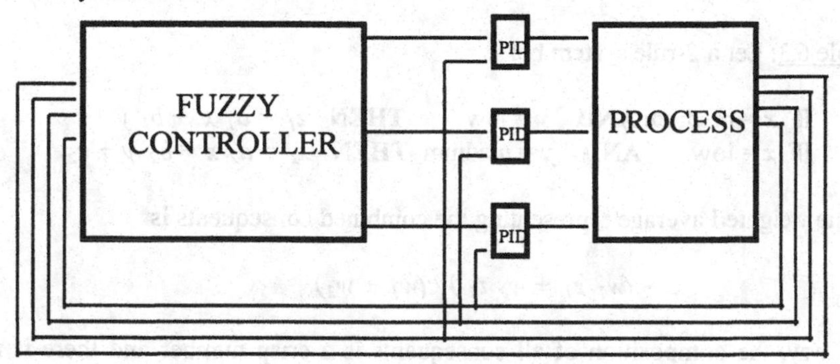

Figure 6.9. Fuzzy supervisory control

6.13 Parametric fuzzy controllers

The parametric form of fuzzy rules was briefly mentioned in Section 5.2 as the second alternative fuzzy controller structure. The general format of a rule in parametric form is:

$$\textbf{IF } X = A \textbf{ AND } Y = B \textbf{ THEN } Z = f(X,Y) \tag{6.17}$$

where A and B are fuzzy sets of the antecedent and $Z = f(X,Y)$ is a crisp function of the consequent. The function $f(X,Y)$ could be any function that describes the system output within the fuzzy region of the control space defined by antecedents X and Y.

When $f(X,Y) = constant$, we speak of a *zero-order* parametric structure where the consequent is a fuzzy singleton or a fuzzy membership function, as in a rule-based structure. When $f(X,Y)$ is a *first-order* polynomial, we have

$$\textbf{IF } s_1 = S_1^i \textbf{ AND } s_2 = S_2^i \textbf{ THEN } Z^i = a_0^i + a_1^i s_1 + a_2^i s_2 + \ldots + a_p^i s_p^i \qquad (6.18)$$

where s_i is an input variable; Z^i an output variable, S_j^i is a linguistic fuzzy membership function, and the coefficient set $\{a_j^i\}$ is the parameter set to be identified.

The principles of this systems identification algorithm is illustrated in Figure 6.10. The approach combines a global rule-base description with *local linear approximations* by means of a linear regression model corresponding to a linear input-output model that would describe the system locally. In other words, it is a hybrid approach that combines fuzzy rule-based antecedents (**IF** part) that define the overlapping input regions and the consequent (**THEN** part) that define linear approximations for the corresponding regions.

As regards combining fuzzy sets within one rule, aggregation proceeds as in a rule-based fuzzy system. However, the composition process (i.e. combining the crisp outputs from the polynomials after substituting the coefficients determined as shown below as well as crisp values of the input variables x and y), consists of taking the weighted average of the consequents of each rule.

<u>Example 6.3</u>: Let a *2*-rule system be:

$$\begin{array}{lll} \textbf{IF } x = \text{medium } \textbf{AND} & y = \text{low} & \textbf{THEN } z_1 = a_1 x + b_1 y + c_1 \\ \textbf{IF } x = \text{low} \quad \textbf{AND} & y = \text{medium} & \textbf{THEN } z_2 = a_2 x + b_2 y + c_2 . \end{array}$$

Then the weighted average representing the combined consequents is

$$z = (w_1 z_1 + w_2 z_2)/(w_1 + w_2).$$

In this way, the composition of all consequents is a crisp number and there is no need for defuzzification. (It is obvious that because of the inherent lack of defuzzification the parametric method would not work in cases where a fuzzy output, such as that mentioned in Note 2 of Chapter 3 would be required). In Example 6.3 the consequents are linear functions of the input variables x and y, and the parameters a_i and b_i are constant coefficients which can be determined by linear multi-regression analysis.

Consider the *2*-rule system of Example 6.3. Assume that a table of values, gained from measurements, is available for a two-input (x and y) single-output (z) system (Table 6.2). The task is to find linear segments of the output function by fitting a straight line on those of its values which correspond to the fuzzy inputs defined by linguistic membership functions:

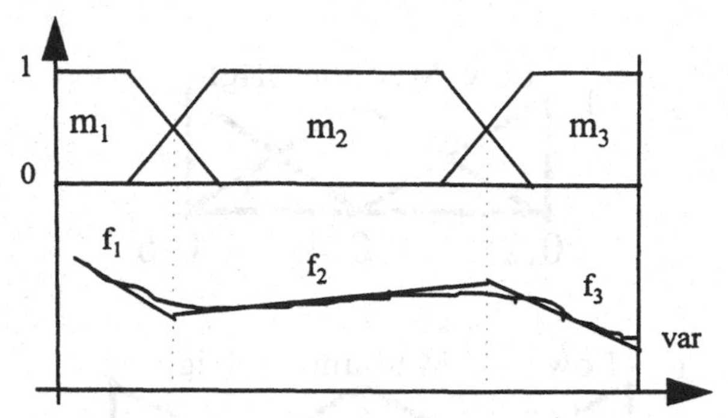

Figure 6.10. Local linear descriptions.

Table 6.2 Input-output table for Example 6.3

	x	*y*	*z*
1	*0.1*	*0.8*	*-0.3*
2	*0.2*	*0.7*	*0.4*
3	*0.15*	*0.6*	*0.7*
4	*0.3*	*0.7*	*0.8*
5	*0.4*	*0.1*	*0.1*
6	*0.5*	*0.5*	*0.5*
7	*0.5*	*0.9*	*-0.2*

Consider Rule 1 above:

IF x = medium **AND** y = low **THEN** $z_1 = a_1 x + b_1 y + c_1$

Assume the above set of membership functions for each input variable x and y. The apex points of membership function *medium* on the horizontal axis for variable x are: *0.2* , *0.6* and those of *low* for variable y are: *0.5* , *0.8*. Go to the table and find those rows where the values of x falls between *0.2* and *0.6* and the values of y fall between *0.5* and *0.8*. These are: Rows *2,3,4*, and *6*. The task is to find a straight line which could be fitted on these points such that the squares of the distances between the line and the points would be minimized. The coefficients a_1 , b_1 , c_1 , calculated by means of a linear regression program package, are the least square coefficients of the straight line fitted of these points. Rule 2 can be processed in the same way, until both rules can be written in terms of the known linear coefficients. Thus, in a way, the linear equation coefficients are *trained* by the example data. *This is analogous to the training phase of a neural network* where example input data and the corresponding desired outputs are presented to the network (refer to Chapter 9). In turn, the crisp consequents can be calculated for various values of input variables x and y.

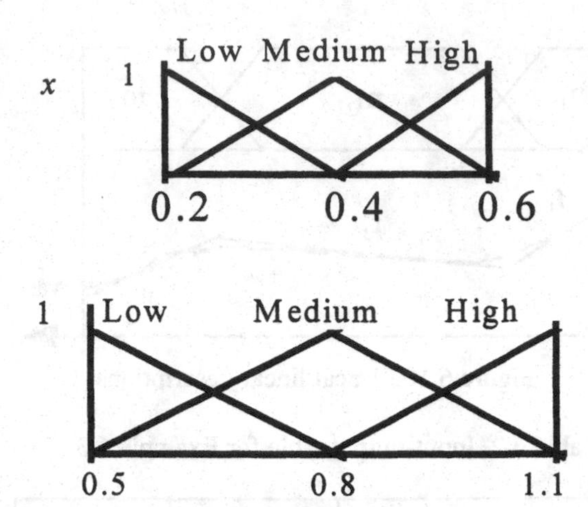

Figure 6.11. Membership functions for Example 6.3

Table 6.3. Input-output table of selected values for Example 6.3

	x	*y*	*z*
2	*0.2*	*0.7*	*0.4*
3	*0.15*	*0.6*	*0.7*
4	*0.3*	*0.7*	*0.8*
6	*0.5*	*0.5*	*0.5*

6.14 Comparison between rule based and parametric fuzzy approaches

- The *rule based* fuzzy approach is more suitable for acquiring and implementing expert human operator knowledge, while the parametric fuzzy approach is best used with input/output numerical data, if available.
- The *parametric fuzzy approach* yields a better estimation accuracy because it is a hybrid of rule based fuzzy and mathematical components. The accuracy of the parametric approach is generally superior to the rule based approach for the same number of rules. Although accuracy can be improved by a larger number of rules and membership functions, one of the reasons for using fuzzy logic is the premise that approximate estimation should be sufficient. Besides, the *Fuzzy Approximation Theorem* shows that any function can be approximated by a fuzzy system to any degree of accuracy desired, albeit at the expense of additional computational processing overhead.
- Whenever algorithms are developed in conventional programming languages, the development time of parametric fuzzy systems is shorter than that of the rule based approach. However, modern commercially available rule based fuzzy development software speeds up the development of rule based systems, although it unfortunately does not cater to the design of parametric fuzzy controllers. The reason is that such software is geared to industrial applications with an emphasis on capturing the linguistic know-how

of human operators. More research-oriented software packages, such as *MATLAB™* where the user can augment the software package with special-purpose routines developed by him, are more amenable to the use of parametric fuzzy systems.

- The parametric fuzzy algorithm is inherently adaptive, because the coefficients can be altered during system tuning. A real-time adaptive implementation of the *parametric* approach is feasible by means of a recursive least-square algorithm, which updates the linear coefficients either at a certain fixed frequency or whenever determined by observed conditions.
- In adaptive versions of the *rule-based* approach , changing the rule weights (*Degree of Support*) or the membership functions does make *non-recurrent* adaptation practically achievable. However, commercially available neurofuzzy development software (i.e. a combination of neural and fuzzy techniques) presently does not cater to *recurrent* parameter updating.
- The disadvantage of the parametric fuzzy approach is the *loss of the linguistic formulation of output consequents*. Thus the same applies here as with the relational equation based fuzzy model identification (see Section 6.15), namely, that in an industrial plant/process control environment the rule-based approach where the human operator's experience plays an important role, is still far more attractive.

6.15 Relational equation based fuzzy system

This represents the third fuzzy system structure alternative outlined in Section 5.2. A discrete fuzzy system can be described either as a set of *fuzzy logical rules* or as a set of *fuzzy relational equations*. Mathematically there is no essential difference between the two approaches. In both cases, results are obtained by some fuzzy operations culminating in a fuzzy relation (the *systems identification phase*) which, as a knowledge base, is used in the next step (the *estimating phase*) to generate outputs from input stimuli. In general, fuzzy modeling consists of two distinct operations: *systems identification* and *estimation*. A direct connection between fuzzy rule-based and relational equation-based systems will also be given as an illustration of the commonality of the two approaches.

6.16 Fuzzy system identification

Systems identification implies here the identification of the *structure* and *parameters* of a fuzzy model such that this model has an input-output behavior which approximates that of the dynamic system being modeled. This operation culminates in the derivation of a *fuzzy relation*, based on N samples, which is assumed to be the fuzzy image of the system being modeled. The choice of discrete-time fuzzy model *structure* is usually based on some previous knowledge of the system being modeled.

The derivation of the fuzzy relation from discrete measurements is a *parameter identification* problem which includes the formulation of a *set of control rules* (fuzzy implications) and the generation of *membership functions*. This may be accomplished either by means of an interview with a skilled control operator, as outlined in previous chapters, or automatically by means of the method of relational equations.

6.17 Fuzzy learning

The process of establishing the fuzzy relation may be construed as a *learning process* for the following reasons. If the fuzzy relation is constructed on the basis of human operator interviews, the learning is done by the fuzzy controller designer who deposits the final results of his learning process into the fuzzy relations. If, however, the fuzzy relation is constructed on the basis of relational equations where the input-output data are acquired from on-line measurements, the fuzzy relation is being built up gradually and is recalculated each time a new set of data has been received. As a result, the construction of the fuzzy relation is seen as a *learning process*, i.e. every new piece of information is being incorporated into the structure of the fuzzy relation. In such a case, linearity, time invariance, etc. are of no consequence, because the fuzzy relation is being "customized" for the specific dynamics of the system being modeled. Thus in relational fuzzy modeling, the modeling period when the fuzzy relation is being built up is referred to as the *learning phase*.

6.18 Evaluation: advantages and disadvantages

The systems identification method using relational fuzzy equations offers a *systematic design procedure* for fuzzy model construction, in as much as it avoids the problems associated with the formulation of fuzzy control rules based on a human operator interview. For example, the control functions verbalized by the skilled human operator are often rather sparse. It is most likely that additional rules will have to be added later on to provide for various effects not taken into consideration initially. In other words, the degree of completeness of the rule base is never certain. (This is described in Chapter 10 along with other practical aspects of fuzzy controller design, including the emphasis upon the iterative tuning of the controller during the design procedure). Besides, the interviewing process may require the expertise of industrial psychologists to construct meaningful interviewing questionnaires and to win the cooperation of human control operators; thus in many complex cases, rule-based design is not a trivial process.

As has been mentioned in Chapter 6, the relational equation approach like the parametric approach proposes *to do away with the need for a skilled human operator* and relies on measurements whose results are *learned* by the fuzzy model. In other words, such a *trainable system* is similar to a neural network, as we will see in Chapter 9. Human experience and qualitative judgment would still be utilized in the process of formulating the membership functions. In addition, the rigorous formality afforded by the relational equation based method provides many valuable insights such as, for example, that of a fuzzy automaton, and discoveries of techniques such as, for example, *recursive estimation*[1] applicable in robotics. However, in general controller design the relational equation based fuzzy model identification proceeds along the lines of the classical approach whereby the plant/process itself is being identified (unless, of course, the approach is applied to the identification of a human controller while he/she is controlling the plant/process, as shown in Chapter 2). As a result, the difficulties having to invert the model in order to develop the controller, as discussed in Section 4.20, still remain. The main disadvantage of the approach is that according to its latest research status, the relational equation method is applicable only to single-output systems, yet many industrial systems require a multi-output capability. In a real industrial environment, rule-based fuzzy modeling is still the favored one because it is easier to understand and troubleshoot, it can handle multi-input multi-output systems and it is more in line with the original premise of fuzzy control in that it provides a method to translate human experience into control laws.

6.19 Estimation, defuzzification, and heuristic optimization

The second step of this fuzzy model design procedure is the *estimation* of outputs based on inputs to the fuzzy model such, that its behavior emulates that of the dynamic system being modeled. In the *estimation phase* the fuzzy relation is used along with measured on-line data and its performance is compared with that of the system being modeled. In turn, he estimated model output vector is obtained by means of *defuzzification,* which is a transformation from the fuzzy domain to the discrete domain. Finally, the model fit is improved by a *heuristic optimization procedure* consisting of the empirical adjustment of time delays in the discrete fuzzy model. The best combination of these heuristic parameters yields a minimum performance index J which is the variance of the modeling error. The next section begins with the formal definition of a number of fuzzy modeling concepts. In turn, the probabilistic relation-building method is presented, followed by an exposition of fuzzy estimation methods. Finally, a short discussion of heuristic model optimization is given and the connection between rule-based and relational equation based fuzzy systems is shown.

6.20 Fuzzy modeling operations

The dynamic system to be modeled is assumed to be a multi-input single-output (*MISO*) prototype system with input terminals $m = 1,...,M$, and a single output terminal. Let the system for which a fuzzy model is to be constructed be

$$S(u_1[k], u_2[k], ..., u_m[k], y[k]) \qquad (6.20)$$

where $u_m[k]$ and $y[k]$ represent time vectors of forced inputs and the output vector respectively, as applied to the m terminals. Furthermore,

$$u_m[k] \in U_m, \; m = 1,...,M; \text{ and } y[k] \in Y \qquad (6.21)$$

where U_m and Y are the particular universes of discourse pertaining to the mth forced input vector and the output vector, and time samples $k = 1,...,Q$.
The fuzzy relation is

$$\Re \subset U_1 \, X \, U_2 \, X \, ... \, X \, U_M \, X \, Y \qquad (6.22)$$

where X symbolizes the Cartesian product. Then the general fuzzy model of this system may be expressed as

$$Y = U_1 \circ U_2 \circ \ldots \circ U_m \circ \ldots \circ U_M \circ \Re \qquad (6.23)$$

where' ° ' is the fuzzy compositional operator. This is an *input-output description* of the fuzzy model. In the following, a fuzzy *state space* model will be developed. In order to take into account the internal dynamics of the system to be modeled, let us introduce

$$x_n[k-1] = y[k-\tau_{yn}], \quad n = 1,\ldots,N \qquad (6.24)$$

as internal states of the system S, which are *delayed* values of the output $y[k]$, $k = 1,\ldots,Q$. (If incorporated in an input-output description, these may be thought of as *initial conditions*, i.e. system inputs at $k = 0$.) These values can also be expressed in terms of an internal state time vector consisting of the N internal states $x_n[k-1]$; furthermore $x_n[k-1] \in X_n$, $n = 1,\ldots,N$.
The *forced input* time vector is defined as

$$u_m[k] = u_m[k-\tau_{um}], \; m = 1,\ldots,M \qquad (6.25)$$

Assume that the desired fuzzy model's structure is of the Nth order with M external inputs and a single output:

$$y[k] = (u_1[k] \circ u_2[k] \circ \ldots \circ u_M[k]) \circ (x_1[k-1] \circ x_2[k-1] \circ \ldots \circ x_N[k-1]) \circ \Re \qquad (6.26)$$

where $\Re$ is the estimate of the fuzzy relation.

Definition 1. Given a vector z consisting of the real variables z_i over the space Z normalized over the range $-A \ldots +A$.

$$z = \{z_1, z_2, z_3, \ldots, z_i, \ldots, z_j\}, \; z_i \in Z \qquad (6.27)$$

and assume that space Z is partitioned into ρ referential fuzzy sets comprising of isosceles triangles which satisfy the requirements of being *normal, convex* and *completely covering the space Z, and partially overlap*, as in Figure 6.12. Then for every component value z_i of the vector z plotted on the horizontal axis, the ρ referential fuzzy sets generate a fuzzy variable z_i' which is a ρ-term set:

$$z' = \{z_{i1}', z_{i2}', z_{i3}', \ldots, z_{ir}', \ldots, z_{i\rho}\} \qquad '(6.28)$$

For example, in Figure 6.12, $\rho = 5$ and the real value indicated on the horizontal axis yields the 5-term set $\{0, 0.8, 0.2, 0, 0\}$.

Expressing the ρ referential fuzzy sets in terms of their *membership functions* μ_r, $r = 1,\ldots,\rho$, yields

$$z_i' = \{\mu_1(z_i), \mu_2(z_i), \ldots, \mu_r(z_i), \ldots, \mu_\rho(z_i)\} \qquad (6.29)$$

where z_i is one particular value of the real variable.

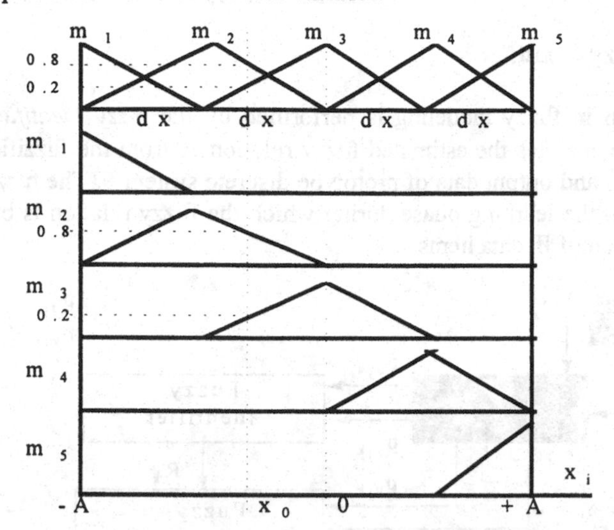

Figure 6.12. Membership functions of reference fuzzy sets.

Definition 2. Since each real-number component of a J-element vector z can give rise to a fuzzy ρ-term set, the vector z gives rise to J fuzzy ρ-term sets or, in other words, ρ fuzzy vectors of J components each. The ***possibility vector*** refers to ρ vectors whose components have identically indexed μ membership functions. That is, the possibility vectors for $z = \{z_i\}$, $i = 1,...,J$ are

$$z_i' = \{\mu_r(z_i)\}, \quad i = 1,...,J, \; r = 1,..., \rho \tag{6.30}$$

Definition 3. Using the fuzzy partition of space Z into ρ referential fuzzy sets, *fuzzification* is defined as the *1-to-*ρ mapping whereby a real value is converted into ρ fuzzy values. We define a *fuzzifier operator* Φ_ρ over ρ referential fuzzy sets as

$$z_i' = \Phi_\rho(z_i) \tag{6.31}$$

Definition 4. Using the fuzzy partition of space Z into ρ referential fuzzy sets, *defuzzification* is defined as the ρ*-to-1* mapping whereby the fuzzy ρ-term set is converted into a single real value. A *defuzzifier operator*, Φ_ρ^{-1} is defined over ρ referential fuzzy sets as

$$z_i = \Phi_\rho^{-1}(z_i') \tag{6.32}$$

The defuzzifier operator may be defined as

$$z_i = \Phi_\rho^{-1}[\,\Phi_\rho(z_i)] \tag{6.33}$$

or by implication,

$$z_i' = \Phi_\rho[\Phi_\rho^{-1}(z_i')] \tag{6.34}$$

6.20.1 Fuzzy identifier

The first step in fuzzy modeling is performed by the *fuzzy identifier* (see Figure 6.13) which generates the estimated fuzzy relation $\mathfrak{R}$ from the fuzzified input data, internal states and output data of prototype discrete system $\boldsymbol{S}$. The fuzzy identifier is active only in the learning phase during which the fuzzy relation is built up on the basis of a batch of W data items.

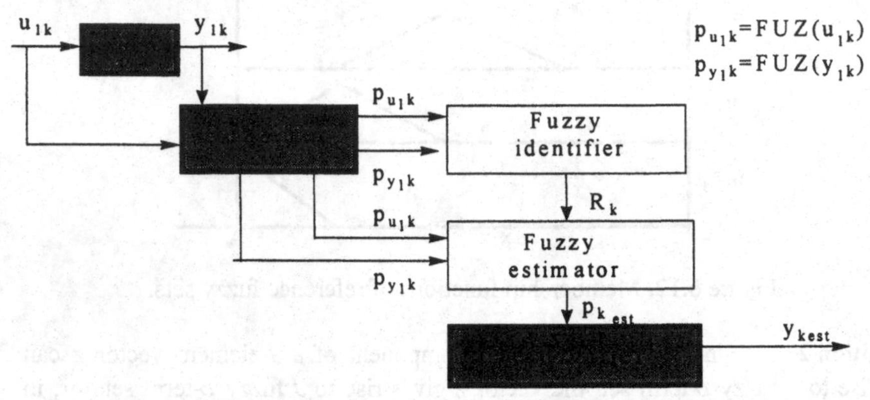

Figure 6.13. Relational equation based fuzzy system design steps.

Definition 5. The general form of the estimated fuzzy relation $\mathfrak{R}$ generated by a fuzzy identifier is defined as

$$\mathfrak{R}(U_m, X_n, Y) = \Phi_\rho(\boldsymbol{u}_m[k]) \circ \Phi_\rho(\boldsymbol{x}_n[k-1]) \circ \Phi_\rho(y[k]) \tag{6.35}$$

where $m = 1,\ldots,M$, $n = 1,\ldots,N$, $k = 1,\ldots,L$. The fuzzy identifier discussed here utilizes the so-called *probabilistic fuzzy relation-building approach* whose essential feature is to regard the possibility vector as being what its name implies, that is, a measure of possibilities of a variable lying in the ρ referential fuzzy sets. An entry in the fuzzy relation matrix measures the possibility of obtaining an output $y[k]$ in space Y from forced inputs $\boldsymbol{u}_m[k-\tau_u]$ and internal states $\boldsymbol{x}_n[k-\tau_y]$ in spaces U_m and X_n, $m = 1,\ldots,M$; $n = 1,\ldots,N$, respectively. To estimate this on the basis of past experience, the weighted average

$$\mathfrak{R}(U_m, X_n, Y) = \Sigma\, \varphi_u\, \varphi_x \,.\, y'[k] \,/\, \Sigma\, \varphi_u\, \varphi_x \tag{6.36}$$

is taken, where $\varphi_u\, \varphi_x$ is the product $\boldsymbol{u}_m[k-\tau_u] \,.\, \boldsymbol{x}_n[k-\tau_y]$, $m = 1,\ldots,M$, $n = 1,\ldots, N$, the possibilities that the individual inputs belong to the particular reference sets. This product can be interpreted as the **AND** combination of a particular group of input reference sets and represents the possibility that the actual input can be represented by this group of sets. The summation runs over the relevant observations k, $k = 1,\ldots,Q$. This is analogous to estimating a *probability* from a table of frequencies of occurrence, and the array $\Sigma\, \varphi_u\, \varphi_x$ can be thought of as measuring the frequencies of occurrence of the various combinations of inputs and internal states. Postlethwaite found that this *probabilistic* fuzzy identifier is especially suitable to industrial systems. because of its superior noise rejection capabilities due to the averaging effect discernible in Equation 6.36.

6.20.2 Fuzzy estimator

The second step of fuzzy modeling is performed by the *fuzzy estimator* which uses the fuzzy relation $\Re$, along with the fuzzy variables derived from M forced inputs and N internal states. During the estimating phase the fuzzy model generates an estimate of the output $y[k]$ of system S for $k = 1,...,Q$. The fuzzy model presently discussed has a first-order structure:

$$y'[k] = y_1'[k-\tau_y] \circ u_1'[k-\tau_u] \circ \Re \qquad (6.37)$$

This may immediately be recognized as the *compositional rule of inference*. Figure 6.13 shows the position of the fuzzy estimator in the block diagram that depicts the procedure of computing a fuzzy controller. The fuzzy estimator has the same three inputs as in Equation 6.37.In the probabilistic approach to fuzzy modeling presented here, the fuzzy relation is to be considered as being analogous to the conditional probability of an output, given the combined probabilities of all inputs and internal states, that is,

$$\Re(U_m, X_n, Y) = P\,(Y_m \,/\, U_m, X_n) \qquad (6.38)$$

Since the fuzzy relation $\Re$ measures the possibilities of all input-output combinations, it follows that the expected output $y'[k]$ at sampling time k is the matrix product of the known input and the relation $\Re$. In the present notation and for the first-order case,

$$y'[k] = \sum_{r_u=1}^{\rho} \sum_{r_y=1}^{\rho} \prod_{\substack{r_u=1 \\ r_y=1}}^{\rho} [\mathbf{u}'_{1k}(r_u), \mathbf{x}'_{1,k-1}(r_y), \Re(U_m, X_n, Y)\,] \qquad (6.39)$$

where $\mathbf{u}'_{1k} = \Phi\rho\,(u_1[k-\tau_u),\ \mathbf{x}'_{1,k-1} = \Phi\rho\,(y[k-\tau_y],\ r_u = r_y = 1,...,\rho,\ \rho$ = number of referential fuzzy sets, and $k = 1,...,Q$. Clearly, this fuzzy estimator is limited to *smoothing*, because all observations used for building the fuzzy relation during identification were used in estimation. In other words, the same data that had been used in building the fuzzy relation have been estimated. Another kind of estimator uses a moving average approach which takes into account only the current observation and the past batch of W observations: u_i; $i = k - W, k - W + 1,...,k$. This type of identifier acquires data in the interval $[k - W, k]$ and as the batch W moves forward with each new sample, the oldest sample is rejected. Assume that there is a total of Q data points and $W < Q$. Then it is possible to build the fuzzy relation from only W data and estimate $Q - W$. Finally, a *recursive estimator*[1] may also be used whereby the next fuzzy state is estimated from the previous one, thus it represents a *fuzzy automaton*. This approach extends the input-output formulation of the compositional rule of inference to a set of *N*th-order state space equations. These systems are described in detail in the references by Shaw[1992]. Defuzzification methods were described in detail in previous chapters.

6.20.3 Heuristic optimization

Following defuzzification, the model fit may be optimized by means of a heuristic feedback loop shown for the first-order system in Figure 6.13 and containing heuristic parameters τ_u and τ_y. The best combination of these heuristic parameters yields a minimum performance index J, or modeling error variance, defined as

$$J = [1 / (n_1 - n_2)] \sum_{k=n_1}^{n_2} (y[k] - y\hat{}[k])^2 \tag{6.40}$$

where n_1 and n_2 are defined according to the specific region during which identification or estimation with the given discrete system's behavior take place, and $y\hat{}[k]$ is as defined in the previous section. Thus

$$\tau_u = \tau_u [J(y[k], y\hat{}[k]]; \ \tau_y [J(y[k], y\hat{}[k]]; \ k = n_1, \ldots, n_2. \tag{6.41}$$

The τ delays are related to the sampling rate of inputs and outputs of system $\boldsymbol{S}$ to be modeled. The heuristic feedback loop tends to capture the internal dynamics of S by adjusting the model order as well as the sampling rate.

6.21 Connection between rule-based and relational equation based fuzzy systems

Assume a first-order fuzzy system :

$$p[k] = p_1^{r1}[k] \circ p_2^{r2}[k] \circ \Re \tag{6.42}$$

This is the *compositional rule of inference* where $p_1^{r1}[k]$ is the possibility vector of the forced input, $p_2^{r2}[k]$ is the possibility vector of the internal system states (initial conditions), and $\Re$ is the fuzzy relation. The number of referential fuzzy sets (membership functions) is r_1 for universe of discourse U_1 , r_2 for universe of discourse U_2 , and r for universe of discourse Y. Then the fuzzy relation has the form of an r_1 X r_2 X r matrix of possibility measures, where each entry $\Re_r (U_i, U_j, Y_k)$ measures the possibility p_{ijk} (a number in $[0,1]$) of obtaining an output y in universe of discourse Y from inputs in universes U_i. . For example, if each universe of discourse has been partitioned into 5 overlapping isosceles triangular referential fuzzy sets (membership functions), then the size of the relation matrix is $5 X 5 X 5 = 125$ possible matrix entries. The fuzzy relation $\Re$ is a relation between the referential fuzzy sets defined on each universe of discourse, and is equivalent to the fuzzy rule set in terms of these referential fuzzy sets. Each matrix element can be interpreted as the description of a rule of the following form:

$$\textbf{IF } U_i \textbf{ AND } U_j \textbf{ THEN } Y_k \text{ with possibility } p_{ijk} \tag{6.43}$$

or, in terms of possibility vectors:

$$\textbf{IF } \{p_1^{r1}[k] \textbf{ AND } p_2^{r2}[k]\} \textbf{ THEN } \{p^r[k] = \Re\{r_1, r_2, r\} \tag{6.44}$$

This clearly shows the connection between rule-based and relational equation based fuzzy systems.

Notes

1. See Note 1, Chapter 12 on recursive estimators.

References

BERNARD, JA:"Use of a Rule-Based System for Process Control." IEEE Control Systems Magazine, October 1988;3-13.
CASTRO JL: "Fuzzy Logic Controllers Are Universal Approximators".IEEE Trans.Sys.,Man,Cybern.,1995;25;4;629-635.
DE NEYER M, GOREZ R:"Integral Actions in Fuzzy Control." EUFIT'93 Conference, 1993;156-162, Aachen, Germany.
DUBOIS D, PRADE H: *Fuzzy Sets and Systems:Theory and Applications.* Academic Press, 1980.
FuzzyTECH 3.1 *Explorer Manual and Reference Book,* Revision 310, Dec 1993.
GERRY, J.: "How to Control Processes with Large Deadtimes. Control Engineering, March 1998.
GALLCHET S, FOULLOY L:"Fuzzy Equivalence of Classical Controllers." First European Congress on Fuzzy and Intelligent Technologies (EUFIT '93), Aachen, 1993;, 1567-1573.
HARRIS CJ, MOORE CG, BROWN M: *Intelligent Control: Aspects of Fuzzy Logic and Neural Nets.* World Scientific Publ., 1993, Singapore, ISBN 981-02-1042-6.
JANG JSR, SUN CT:"Neurofuzzy Modeling and Control."Proc.IEEE, 1995;83;3;388-406
KOSKO B: *Neural Networks and Fuzzy Systems.* Prentice Hall, 1992.
KRUGER JJ, ALBERTS HA:"Fuzzy Human-Machine Collaborative Control Of a Nonlinear Plant."IFAC World Congress,1992:8;337-342.
LEE J:"A Velocity Type Fuzzy Logic Controller With Intelligent Integrator." EUFIT'93 Conference, 1993;856-860, Aachen, Germany.
LI YF, LAU CC:"Development of Fuzzy Algorithms For Servo Systems."IEEE C ontrol Systems Magazine, April 1989;65-72.
MAMDANI, EH: "Twenty Years of Fuzzy Control: Experiences Gained and Lessons Learnt" Proceedings of the 2nd IEEE International Conference on Fuzzy Systems, 1993; ISBN 0-7803-0614-5;339-44.
MENDEL JM:"Fuzzy Logic Systems for Engineering: a Tutorial." Proc.of theIEEE,1995;83;3;377.
PEDRYCZ W: "Numerical and Applicational Aspects Of Fuzzy Relational Equations." Fuzzy Sets Syst., 1983;11;1-18.
PEDRYCZ W: *Fuzzy Control and Fuzzy Systems.* John Wiley & Sons, 1989.
POSTLETHWAITE B: "A Model-Based Fuzzy Controller." Trans.I.Chem.E, 1994; 72, Part A; 39-46.
POSTLETHWAITE B: "Empirical Comparison Of Methods Of Fuzzy Relational Identification." IEE Proceedings-D, 1991; 138; 199-206.
RAYMOND C,BOVERIE S, LE QUELLEC JM:"Practical Realization of Fuzzy Controllers :Comparison With Conventional Methods." EUFIT'93 Conference, 1993;149-169, Aachen, Germany.
RIDLEY JN, SHAW IS, KRUGER JJ: "Probabilistic Fuzzy Model for Dynamic Systems." Electronics Letters, 1988; 24; 14; 890-892.
RIDLEY JN: "Fuzzy Set Theory and Prediction of Dynamic Systems." Fulcrum, 1988;18; 1 - 5. Univ. of the Witwatersrand , Johannesburg, South Africa.

SHAW IS, KRUGER JJ: "New Fuzzy Learning Model with Recursive Estimation for Dynamic Systems." Fuzzy Sets and Systems, 1992; 48; 217-229

SHAW IS: Fuzzy Logic - An Introduction. Elektron, July 1992; 8-12.

SHAW IS, KRUGER JJ: "New Approach to Fuzzy Learning in Dynamic Systems." Electron.Lettrs, 1989; 12; 25; 796-797.

SHAW IS: "Fuzzy Model of a Human Control Operator in a Compensatory Tracking Loop." Int. Journal of Man-Mach. Studies, 1993; 39,; 305-332.

SHAW IS: "Fuzzy Logic - An Introduction." Elektron, July 1992; 8-12.

SIMOES MG, BOSE BK:"Application of Fuzzy Logic In the Estimation Of Power Electronic Waveforms." Industry Applications Society Annual Meeting (IEEE-IAS), Vol. II, pp. 853-861, Toronto, Canada,October, 1993.

SUGENO M:"An Introductory Survey of Fuzzy Control." Info.Sciences, 1985;36;59-83.

TAGAKI T, SUGENO M: "Fuzzy Identification Of Systems And its Applications to Modeling and Control." IEEE Trans. Sys.Man.Cybern., 1985, Vol 15, No 1, pp 116-132.

TONG RM: "The Construction and Evaluation of Fuzzy Models." In: *Advances in Fuzzy Set Theory and Applications*, Gupta MM, Ragade RK, Yager RR Eds. North Holland Publ.Co. 1979.

TSAFESTAS S, PAPANIKOLOPOULOS NP:"Incremental Fuzzy Expert PID Control".IEEE Trans.on Industr. Electronics,1990;37;5;365-371.

VERBRUGGEN HB, BABUSKA R: "An Overview Of Fuzzy Modeling for Control." Control Eng.Practice,1996;4;11;1593-1606.

XU CW: "Fuzzy Systems Identification." IEE proceedings, 1989;136;4;146-150.

XU CW, LU YZ: "Fuzzy Model Identification and Self-Learning for Dynamic Systems." IEEE Trans.Sys.Man.Cybern., 1987; 17; 4; 683-689.

7 SYSTEM IDENTIFICATION FOR RULE-BASED SYSTEMS

7.1 Where do fuzzy rules and membership functions come from?

Rule-based fuzzy systems translate qualitative, vague and imprecisely formulated human experience and judgment into control rules. The first section presents the technique to capture this experience. As has been said in Chapter 2, it is the human operator that is being identified while he is controlling the plant/process, thus the rules of the fuzzy control algorithm will constitute an inverse of the plant/process input-output relationship. Membership functions are determined empirically, on the basis of given guidelines, by trial-and-error. In another case, the control system designer himself is called upon to act as a human operator and formulate the rules on the basis of his engineering judgment rather than previous experience with the system. He still uses fuzzy rules and membership functions and the inverse aspects of human control still prevail. As before, membership functions are determined empirically, on the basis of given guidelines, or by trial-and-error. A further variety of the previous methods occur when the variables are *inherently fuzzy,* i.e. cannot be quantified in any way. This aspect affects mostly the choice of membership functions which, in this case, carry the bulk of the empirical knowledge of the human operator or designer, while the rules are established as before. In addition, a brief overview of methods that avoid the use of human operator interviews and "automatically" generate rules and/or membership functions are presented for the case when no experienced human operators are available. In this case, the rules and/or membership functions are *learned* from on-line measurements. The techniques presented are as follows:

- Human operator interview
- Engineering judgment
- Inherently fuzzy variables
- Brief overview of learning systems:
  - Fuzzy self-learning
  - Neurofuzzy system

7.1.1 Human operator interview

Conventional controller performance is often specified in terms of the closed-loop steady-state behavior (steady state error) and the transient response (i.e. unit step response) of a feedback control system. Equally, fuzzy controller rules can also be specified by means of the desired steady state and transient responses expressed in fuzzy terms such as "fast rise time", "minimum overshoot", and "almost zero steady state error". The following example by Harris, Moore and Brown [1993]considered the open-loop step response of a second-order process (Figure 7.1) which requires a closed-loop response with a *fast rise time* and *minimum overshoot*. Assume that the input variables are the system error, e, and the error change, de and that the output variable is u. Assume further that the fuzzy membership sets U, E, and ΔE are represented by seven linguistic fuzzy qualifiers *NB,NM,NS,AZ,PS,PM,PB* with membership functions shown in Table 7.2. Hence there are $7 \times 7 = 49$ possible combinations of the antecedents generating a possible *49* rules of the general form:

$$R_n\text{: } \textbf{IF } e = E^r \textbf{ AND } de = \Delta E^r \textbf{ THEN } U^r \tag{7.1}$$

where $n = 1,2,...,N$ = number of rules, $r = 7$. When the rules are connected with the union operator **OR**, the combined fuzzy control is expressed as:

$$R = R_1 \cup R_2 ... \cup R_N = \bigcup_{n=1}^{N} (E^r X \ \Delta E^r X \ U^r) \tag{7.2}$$

Using *max-min* composition, the membership function of the fuzzy relation has the form:

$$\mu_R(e_i, de_j, u_k) = max\ [min\ (\mu_E(e_i), \mu_{\Delta E}(de_j), \mu_U(u_k))] \tag{7.3}$$

The controller output is obtained from the compositional rule of inference:

$$U = (E\ X\ \Delta E) \circ \Re \tag{7.4}$$

and in terms of membership functions:

$$\mu_U(u) = max\ [min\ (\mu_E(e_i), \mu_{\Delta E}(de_j),\ \mu_R(e_i, de_j, u_k)\ )] \tag{7.5}$$

An experienced human operator would consider the crossover and max/min points of the open-loop system response (Figure 7.1) and *suggest* appropriate control actions at each point to generate a closed-loop response with fast rise time and minimum overshoot. The definitions of linguistic membership functions for a few entries are given in Figure 7.2 and the operator's suggested control actions are inserted into the partial *fuzzy rule map* shown in Figure 7.3.

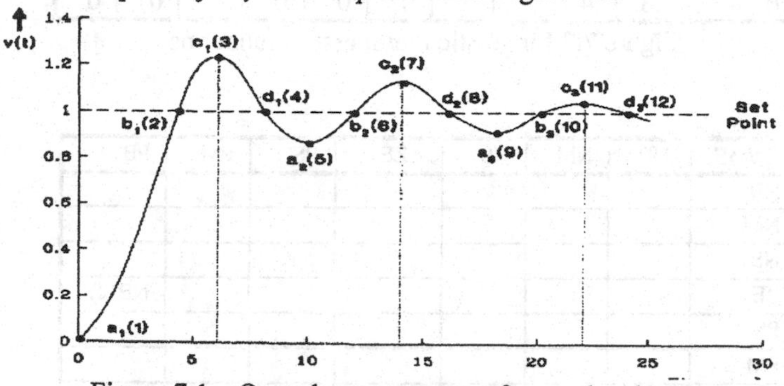

Figure 7.1 Open-loop response of second-order process

For example, at point $a_1(1)$, $E = PB$ for fast rise time ($PB = 5$), but $\Delta E = AZ$ ($AZ = 0$ at $PB = 5$, see Table 7.1). Thus a large control action PB is required to drive the closed-loop response toward the set point with a fast rise time. These values are entered into the rule map of Figure 7.2. On the other hand, at $b_1(2)$, $E = AZ$ and $\Delta E = PB$ and to prevent a large overshoot, control action NB has been used. In this way, the entire rule map can be readily generated (only a partial rule map is shown here).The number in parenthesis represents the control rule number, i.e. $c_2(7)$ point in Figure 7.1 produces the 7th rule with a NM control action. The defuzzified controller response, trace (I), is indicated in Figure 7.4. The response is well-damped but rather slow, since in the example only *13* of the possible *49* rules have been utilized (or alternatively, there are *36* combinations of $(E_i, \Delta E_i)$ for which there is no control decision shown). Better control can be achieved by defining more rules (trace II, Figure 7.4) or using finer partitioned subspaces, although more than seven membership functions are hardly ever required. Besides, it is questionable whether or not the human operator being interviewed can interpolate any finer between points. This example illustrates the apparent coarseness of converting human operator knowledge directly to control rules. However, as has been shown in Chapter 6, the overlapping membership functions tend to make the input-output surface smooth. In these calculations, no fuzzy software development system with a *3*-dimensional input-output display facility was used. This again points out the importance of using the right software tools for practical designs.At any rate, the above procedure has only been used to illustrate a system identification exercise whereby the human operator's fuzzy input-output relationship was identified while he was controlling the plant/process. This fuzzy input-output relationship can also be construed as the inverse of the fuzzy model that drives the plant/process.

	-5	-4	-3	-2	-1	0	1	2	3	4	5
PB	0	0	0	0	0	0	0	.1	.3	.7	1
PM	0	0	0	0	0	.1	.3	.7	1	.6	0
PS	0	0	0	.1	.3	.6	1	.7	.4	.1	0
ZE	0	.1	.3	.6	1	.7	.4	.1	0	0	0
NM	.3	.7	1	.7	.5	.1	0	0	0	0	0
NB	1	.8	.4	0	0	0	0	0	0	0	0

Figure 7.2 Linguistic membership functions

E / ΔE	NB	NM	NS	ZE	PS	PM	PB
NB							
NM							
NS							
ZE							NB(2)
PS							
PM							
PB				PB(1)			

Figure 7.3. Partial fuzzy rule map.

7.1.2 Engineering judgment

As has been mentioned before, here the designer, acting as the control operator, formulates the rules on the basis of his engineering judgment. As an example, one might mention the ***inverted pendulum controller*** which has often been used as a benchmark to illustrate the efficacy of fuzzy control. The inverted pendulum, shown in Figure 7.5, consisting of an upright bar with a pivot mounted on a movable car, is a complex nonlinear system whose control objective is to keep the pendulum in an upright position whenever it begins to fall over on either side. Some researchers even published accounts of having designed complex fuzzy controllers for a similar but double-pivoted pendulum .

The design of a conventional controller for this system is difficult in as much as it requires the real-time solution of a set of nonlinear differential equations in time to make the necessary corrections (i.e. move the car in the proper direction with the appropriate speed). The classical solution is not discussed here, in as much as the purpose of this section is only to illustrate the method whereby to design a rule-based fuzzy controller for this system.

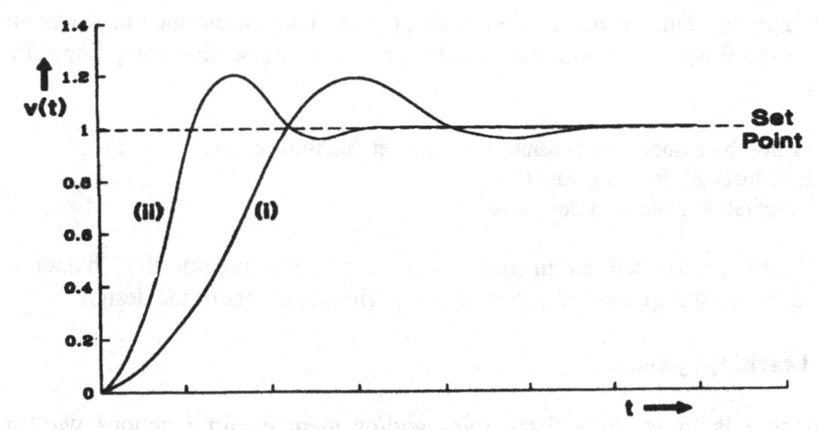

Figure 7.4. Closed-loop response curves

A rule-based fuzzy controller can be designed as follows. The state variables are: pendulum angle ϑ and angular velocity , i.e. change of angle, $\Delta\vartheta$ while the control variable is v, the speed of the car (assuming the positive direction as shown). The designer can reason out the rules, such as, for example:

IF $\vartheta = NM$ *AND* $\Delta\vartheta = ZE$ *THEN* $v = PM$
IF $\vartheta = ZE$ *AND* $\Delta\vartheta = ZE$ *THEN* $v = ZE$, etc.etc.

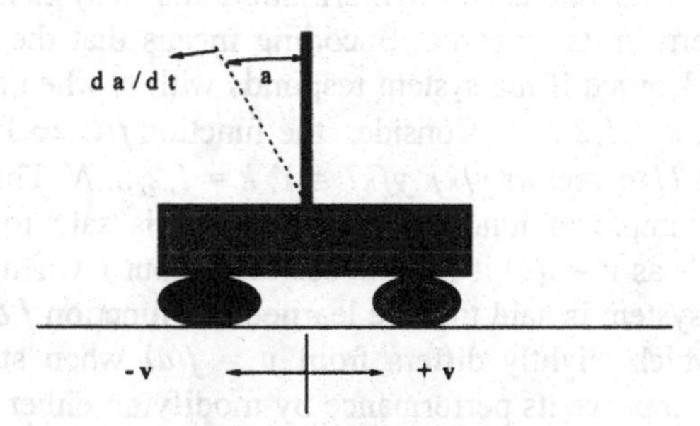

Figure 7.5. Inverted pendulum

In the process of determining these rules the designer actually carries out system identification, in that he establishes the inverse of the inverted pendulum's fuzzy model. It is the inverse because he determines the control actions as if he were a human operator that compensates for the inherent instability of the pendulum.

7.1.3 Inherently fuzzy variables

In some systems, the variables are inherently fuzzy, i.e. no crisp quantitative values can be assigned to them. The designer must assign fuzzy linguistic membership functions to the values assigned to these variables. As an example, let us take the

Japanese fuzzy vacuum cleaner whose design goal is to adjust the suction power on the basis of the following conditional variables and their respective fuzzy linguistic qualifiers:

- floor quality (bare floor, woven bamboo mat, carpet, high-pile carpet)
- amount of dust (this is a crisp variable)
- type of dust (short granular or long wool-like)

The rules are determined as in section 7.1.2 but the membership functions constituted the most important empirical and experiential aspect of the design.

7.2 Learning systems

The goal here is to generate fuzzy rules and/or membership functions without resorting to an experienced human operator who has an intimate knowledge of how to control the plant in question. Ultimately this leads to the *adaptive on-line generation of the fuzzy relation*, i.e. the knowledge base of the fuzzy system. Although the original purpose of using fuzzy logic in control systems was to incorporate the human operator's contribution, such an experienced operator might not always be available. In such cases one must somehow regenerate the operator's knowledge by means of a learning process whereby the sampled input-output plant parameters are measured recorded in terms of input/output time vectors. In another type of application, such as robotics, only a limited number of functions need to be based on learning and these functions are *embedded* in a larger conventional or intelligent controller. *Learning* encodes information, thus a system learns a pattern if it encodes that pattern in its structure. Encoding means that the stimulus-response pair (u_i, y_i) has been learned if the system responds with y_i when stimulated by u_i at every time sample k, $k = 1,2,...,N$. Consider the function $f: U \Rightarrow Y$, i.e. a mapping of vectors $\boldsymbol{u}[k]$, $\boldsymbol{u}[k] \in U$ to vectors $y[k]$, $y[k] \in Y$, $k = 1,2,...,N$. The input-output pair (u_i, y_i) represents a sample of function f. The system is said to have learned the function f *accurately* as $y = f(u)$ if it produces the output y when the input is u. On the other hand, the system is said to have learned the function f *approximately* if it responds with y' which slightly differs from $y = f(u)$ when stimulated by u. A learning system can improve its performance by modifying either its *structure* or its *parameters*. That is, the *knowledge base of a learning system is changed dynamically* so as to keep the knowledge content up-to-date or to improve its quality.

One type of learning takes place in the presence of a performance evaluator which acts as a *supervisor* that decides about the necessary structural or parameter modifications by comparison with a given predetermined standard. If a difference from this standard is detected, an error signal is generated, which is used to reinforce the learning process by "rewarding" accuracy, and "punishing" inaccuracy. Another kind of learning called *unsupervised* or *self-learning* involves no supervision (i.e. no comparison with a preset standard). Such systems look for similar patterns and assemble them into clusters. In fact, this is how biological systems (such as, for example, pre-school children) learn. Two other learning methods that originated from the field of neural networks, are worth mentioning: *linear competitive learning* and *differential competitive learning*. Putting it in a simplified manner, a competitive learning process may be considered as a competition between neurons of a neural network. In a fuzzy system, a similar competition may occur between the rules. The rule that wins the competition has the highest *Degree of Support,* i.e. *it assumes a dominant importance in the rule set.* The difference between the linear and differential approaches is that in the linear approach the rules are adapted continually during the learning process, whereas in the differential approach rules are adapted only if there is a change in the data set. In the linear approach, the weight of a rule (i.e. its *Degree of Support*) is continually modified whenever that specific rule is fired, only to be normalized to unity again. Such unnecessary iterations are avoided by the differential approach. The designation "learning system" implies that the estimation of the output relies on past long-term experience and that *new data items continually modify the fuzzy relation by being incorporated into its structure*. Thus learning also implies *self-organizing* in the sense that the system builds up its own structure on the basis of the data. In this process, old stored patterns are replaced by the most up-to-date pattern that represents the sampled environment. Thus the *rate of forgetting* is as important as the *rate of learning*. The advantage of such a system lies in the fact that it is "customized" according to the kinds of data it is expected to handle, thus it obviates the need for a mathematical model and the concomitant assumptions of linearity and time-invariance. This feature is of particular benefit whenever the plant/process is poorly known.

7.3 Fuzzy learning

One can think of the fuzzy relation as an internal *fuzzy image of the crisp prototype process* from which it was derived. In effect, during the fuzzy model identification process, this fuzzy image is *shaped* or *induced* by the table of input-output data presented to it. In the crisp prototype system, the relationship between the input and output data items generated by a crisp relation R at each sampling time would serve as a *priming function* which would induce some desirable characteristic in the fuzzy relation. In turn, the thus identified fuzzy relation $\Re$ would represent the fuzzy model of the *priming function*. Thus it appears that the priming function would actually "teach" the fuzzy model to assume a particular behavior, hence identifying the fuzzy model with specified characteristics would be tantamount to priming. Once "taught", the fuzzy relation (which is an array of real numbers of order r^{n+1}), connects the input and one output variable y_k, given input u_k . The inputs and output must necessarily be within the universe of discourse (i.e. dynamic range) of the *priming* function. The aforementioned concept of priming qualifies as *self-learning*, in as much as the fuzzy relation is built up step-by-step while the fuzzy image of the currently modeled real-world process is being created.

The fuzzy self-learning discussed is a unique attribute of the relational equation based fuzzy system. It was implied that learning, i.e. the construction of the fuzzy relation, occurs, and thereafter the system carries out fuzzy estimation on the basis of this fuzzy relation and similar learning features, as the next paragraphs will show.

7.4 Rule-based fuzzy learning system

A rule-based fuzzy system can also learn by modifying the rules and the linguistic fuzzy sets used during the learning phase. This procedure also leads to a dynamically changing fuzzy relation, i.e. *knowledge base.*

7.4.1. Adapting the rules

In a rule-based fuzzy learning system rules are assigned a weighting factor: the *Degree of Support* which is a number in the interval *[0,1]* indicating the relative importance of the rule within the set of rules. Since the rule map, i.e. fuzzy relation, mirrors the rule set, the *Degree of Support* is actually a multiplier of each entry in the rule map. Initially all of these weights are set to unity. As the rule map is being dynamically updated, the *Degree of Support* is adjusted as required. Consistent application of this weight management procedure tends to amplify rules that contributed significantly to the fuzzy output and dampens those that contributed less. Thus the system will display a bias towards the areas where most of the data are located. In essence, the adaptive system amplifies the rules that are clustered closest to the most important region of data. The amplification of rule weights means that the system is adapting to a change in its central region of control. This region in a fuzzy system is the middle portion of the controlling fuzzy sets which represent the states of stability or equilibrium of the plant.

7.4.2 Adapting the membership functions

While adjusting the contribution weights of the rules can move the central control region of a fuzzy controller in response to changes in the input, weight modification alone is not sufficient to allow rapid multidimensional adaptation. Such systems must also reorganize their membership functions in response to changes to the changes in the operating environment. The membership functions are adjusted by widening or shifting them slightly, as shown in Figure 7.6. If the system response in the previous execution cycle was above the expected or desired value, then the domain of the membership function accessed is slightly narrowed, that is, the left edge of the domain is moved slightly to the right and the right edge is moved slightly to the left. Similarly, if the system response was below expectation, the domains involved are slightly widened, i.e. their left edges are moved slightly to the left and their right edges are moved slightly to the right. This modification is, however, constrained by a restriction of the amount of *overlap* permitted between adjoining fuzzy sets. According to that restriction, for a vertical line drawn through a region of overlap, the truth membership of the points where the line intersects with the fuzzy sets must always total less than or equal to *1*. If this rule is not adhered to, instabilities can occur. In a dynamically adapting model, the readjustment of neighboring fuzzy regions mans that the overlap limits of the fuzzy membership functions are also continually changing.

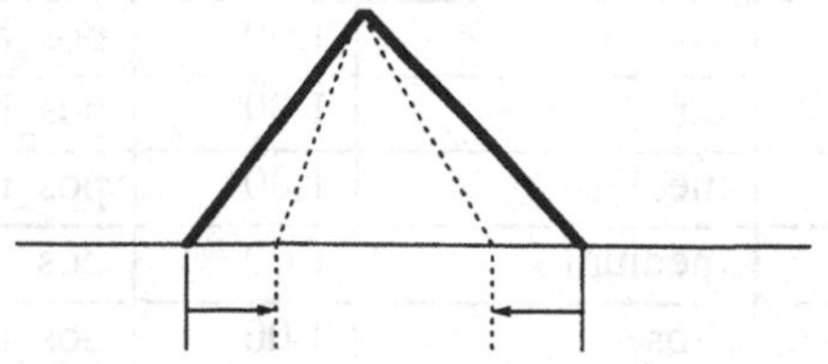

Figure 7.6. Adaptive adjustment of membership function width

7.5 Adaptation mechanism of fuzzy learning systems

An in its time pioneering rule-based adaptive fuzzy controller by Proczyk and Mamdani used supervised learning whereby the controller structure is developed gradually from an empty fuzzy relation on the basis of the data acquired. In the fuzzy controllers used today, however, the adaptation mechanism depends on the basic fuzzy system structure such as rule-based, parametric, or neurofuzzy. In a rule-based fuzzy controller, it involves changing the *Degrees of Support* (i.e. the relative weight of a rule) shown as *DoS* in Figure 7.7 and shifting the end points of the membership functions as in Figure 7.6. This may be carried out by a neural network as in Chapter 9 where the adaptive combination of a neural network and a fuzzy controller is referred to as a *neurofuzzy controller.* It is, however, only a one-time adaptation, at least in the form available as an option in a commercially available fuzzy controller software development package. Chapter 9 discusses the practical use of commercially available neurofuzzy system design software. In a parametric fuzzy controller, a one-time adaptation can be achieved in a manner described in Chapter 6. However, an on-line adaptation, i.e. finding the coefficients of the linear expression in the consequents, is also theoretically possible (see Chapter 13).

7.6 Adaptive rule based fuzzy system using product space clustering

In order to determine how the rules are derived in an adaptive manner, Kosko presented a method based on *product space clustering* as follows. The universes of discourse of input and output state variables of the system are partitioned into a number of sharply defined subfields. These subfields can also be defined as the bases for the referential subfields within the universes of discourse, as shown in Figure 7.8. The combination of a certain input and a certain output pair may be plotted in a matrix. In as much as every matrix element represents a number of input-output combinations that fall into the corresponding input and output subfields, it is reasonable to say that the matrix elements are equivalent to rules. The weight that may be assigned to a rule is then defined as the *frequency of occurrence* of the rule for a number of samples, as shown in Figure 7.8 and in its corresponding histogram, Figure 7.9 while Figure 7.10 shows the distribution of input and output data points within the rule matrix. The darker the color of the matrix entry the more data points have fallen into it.

	IF			THEN
	Angle	Distance	DoS	Power
1	zero	far	1.00	pos_medium
2	neg_small	far	1.00	pos_high
3	neg_small	medium	1.00	pos_high
4	neg_big	medium	1.00	pos_medium
5	pos_small	close	1.00	pos_medium
6	zero	close	1.00	zero
7	neg_small	close	1.00	pos_medium
8	pos_small	zero	1.00	neg_medium
9	zero	zero	1.00	zero

Figure 7.7.Illustration of the Degree of Support.

As can be seen, the rule *IF x = ZE THEN y = ZE* has the highest frequency of occurrence which, as a consequence, will have the largest weight.

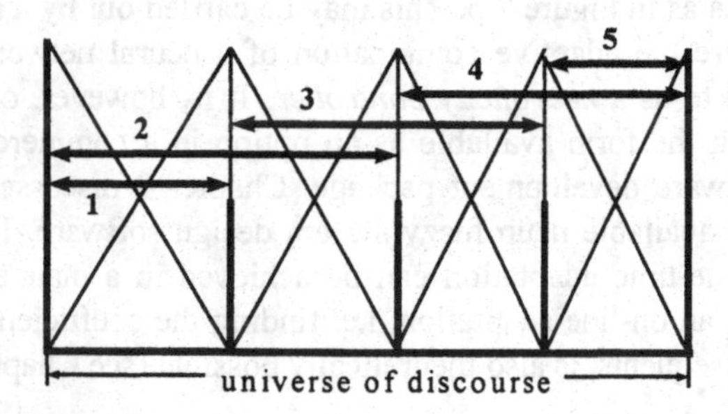

Figure 7.8 Adaptive product space clustering subfields

7.7 The fundamental postulate of predictability

Given a fuzzy relation $R(U, Y^I, Y)$, *primed* during the discrete time interval $1 \leq k \leq M$, $k = 1,2,...,N$ by fuzzified vectors $\boldsymbol{u}'(k-\tau_u) \in U$, $\boldsymbol{y}'(k-\tau_y) \in Y^I$, $\boldsymbol{y}'(k) \in Y$. Any new input data items $u(k-\tau_u)$ received in the discrete time interval $M + 1 \leq k \leq N$, and processed by a predictor, is predictable if and only if $u'(k) \in U$ (which is the same universe of discourse U used for constructing the fuzzy relation). If this condition is not satisfied, clipping (i.e. saturation) might occur in one of the fuzzy model components. In the above formulation, the fuzzy relation is combined from universes of discourse of forced input vector $\boldsymbol{u}'(k)$ delayed by τ_u (this term accounts for delays in the system being modeled), output vector $\boldsymbol{y}'(k)$ delayed by τ_y (i.e. an initial condition vector), and output vector $\boldsymbol{y}'(k)$. The fuzzy relation was built up in the learning (priming) phase while the prediction of outputs to new stimuli was achieved in the estimation or prediction phase. In words, this postulate says that to obtain good predictions the *universe of discourse of such predictions must be the same as that used for building up the fuzzy relation from measured variables.*

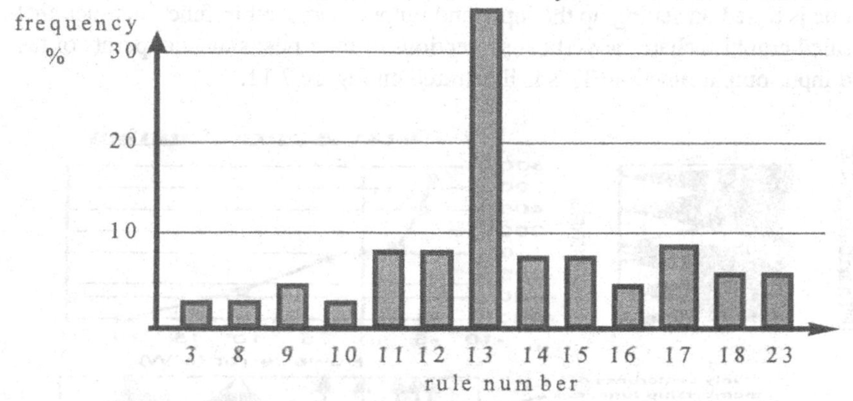

Figure 7.9. Frequency of occurrence of firing a rule

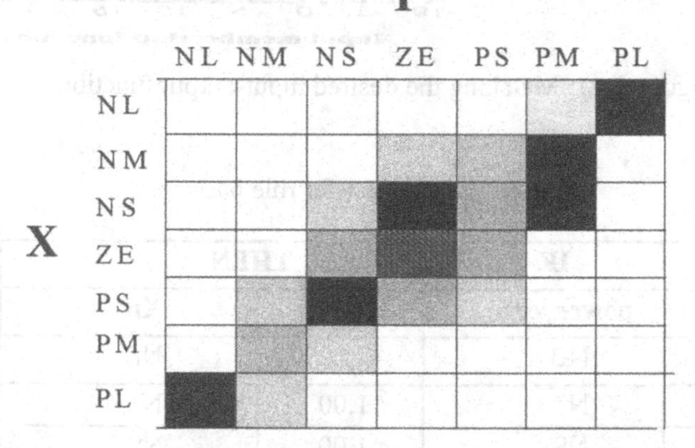

Figure 7.10. The color of a matrix entry (i.e. rule) indicates its weight

7.8 Linguistic synthesis of an input-output function

Fuzzy logic is ideally suited to the linguistic synthesis of an input-output function of a desired form. This is because a general fuzzy logic block has no memory (i.e storage capability) thus its input-output function is just a multidimensional non-linear mapping. The method developed by Chilvers [1996] and discussed below is actually a *graphical solution to the parametric method* shown in Chapter 6. As such, it is easier to implement than the parametric method and *it also makes possible to use the fuzzy controller development system software* highlighted in Chapter 10.

7.8.1 Application to a single-input single-output system

Assume a single-input single output block whose input-output function is defined by a table of corresponding discrete values. Assume further that the universe of discourse of these input and output functions is covered by the same (and also a minimum) number of triangular membership functions. The linguistic synthesis technique is based on setting up the input and output membership functions such that on a scaled graphics chart the vertical projections of their peaks meet at points of the desired input-output function. This is illustrated on Figure 7.11.

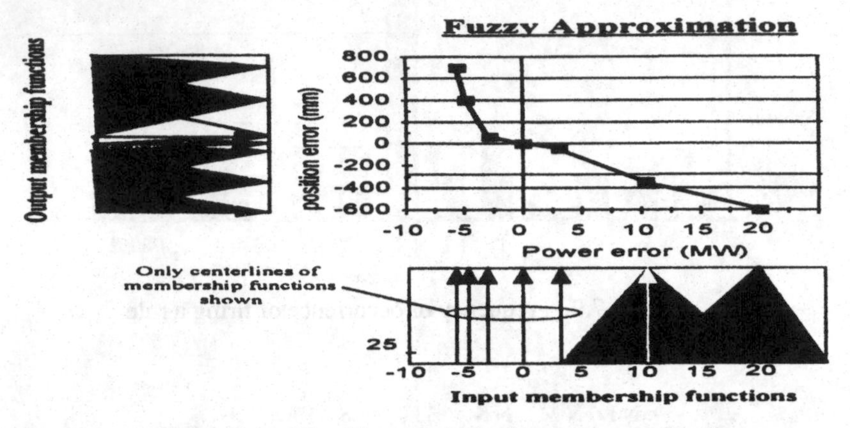

Figure 7.11. Matching the desired input-output function.

Table 7.1. Direct transfer rule base

	IF	THEN	
Rule	power_err	D-o-S	Xr
1	NB	1.00	NB
2	N	1.00	N
3	NS	1.00	NS
4	ZERO	1.00	ZERO
5	PS	1.00	PS
6	P	1.00	P
7	PB	1.00	PB

For a smooth approximation of the desired input-output function, the overlap of the triangular membership functions must be carefully adjusted. In turn, set up the fuzzy rules (for example, by using the *Spreadsheet Editor* of the fuzzy development system of Chapter 10) to map the input directly to the output, in a one-to-one relationship, as shown in Table 7.1. For example, *IF* input = neg_small *THEN* output = neg_small, etc. Clearly, the more inflection points are required, the more membership terms must be used for both the input and output membership functions. This technique can be extended to multidimensional systems such as, for example, the two-input single-output block shown in Figure 7.12, where the input-output function is a surface which shows a good correspondence with the desired function of Figure 7.11.

7.8.2 Application to a two-input single-output system

The example procedure given here describes how to set up a particular desired control surface whose edges correspond to an input-output function of form Y_a ("High Power" in Figure 7.13) when a third variable is $Z = z_p$ and produces an in-

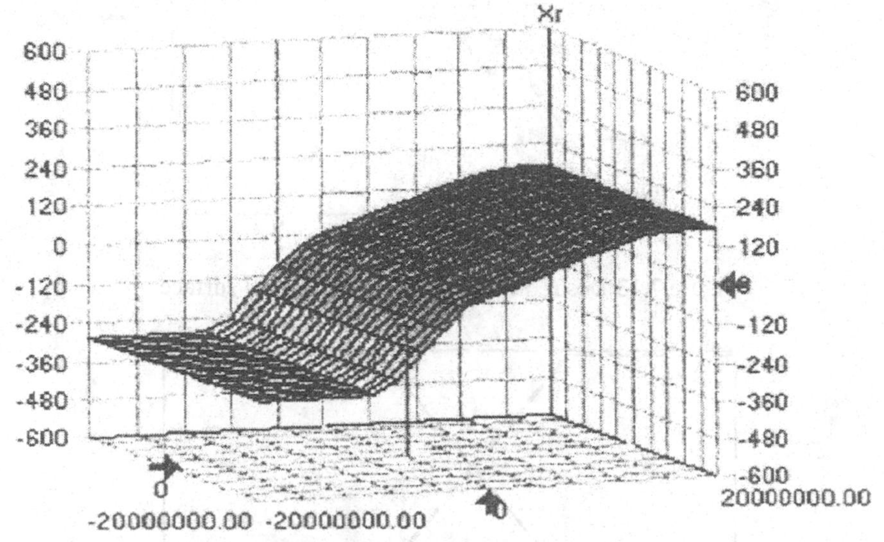

Figure 7.12. Control surface for 2-input 1-output system

put-output function of form Y_b ("Low Power" in Figure 7.13) when the third variable is $Z = z_q$. At values of Z between z_p and z_q the resultant output function is a linear fit between F_a and F_b and forms the third edge of the control surface. (In the control surface of Figure 7.12 the two input functions describing two edges of the surface were identical and the third edge was represented by a constant rather than a linear function). As has been mentioned above, this is the *graphical solution of a parametric fuzzy system* where the condition variables are fuzzy are fuzzy and the consequent variables are linear functions (see Chapter 6). A step-by-step procedure is given below.

- Draw two input-output functions F_a and F_b on a common set of axes, as in Figure 7.15.
- Set up fuzzy membership functions as described in Figure 7.11.
- Set up a third membership function for variable Z as shown in Figure 7.14. For the simplest linear variation of F_a to F_b make this membership function contain two terms.
- For Z as with the other input-output terms, ensure that the membership functions are set up to be evenly overlapping terms which start and end at the same point and cross in the centre. If this is not done, the interpolation will not be smooth.
- Set up the fuzzy rule base to map inputs of Y_a directly to outputs of Y_a but include Z. Repeat this for Y_b but with $Z = z_q$ as shown in Table 7.2. Y_a, but include Z. For example: **IF** input Y = neg_small **AND** $Z = z_p$ **THEN** output Y = neg_small.
- Figure 7.15 shows the control surface achieved in this manner.

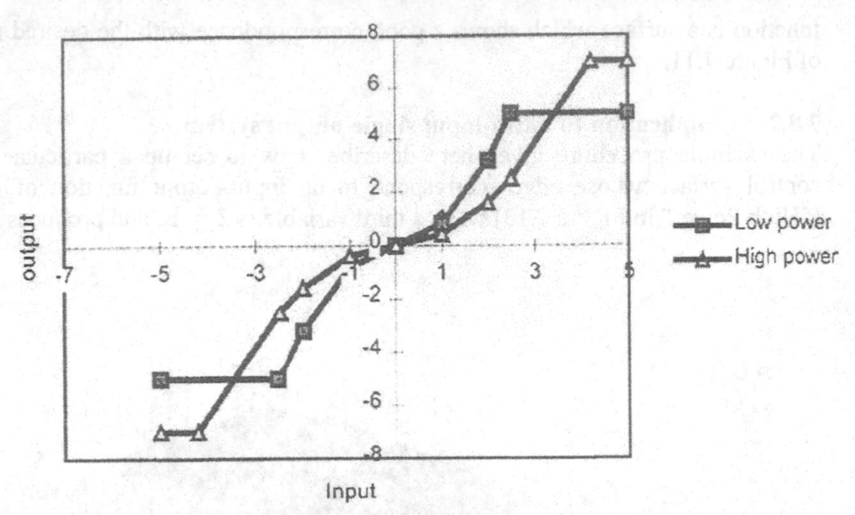

Figure 7.13. Desired edge functions for control surface

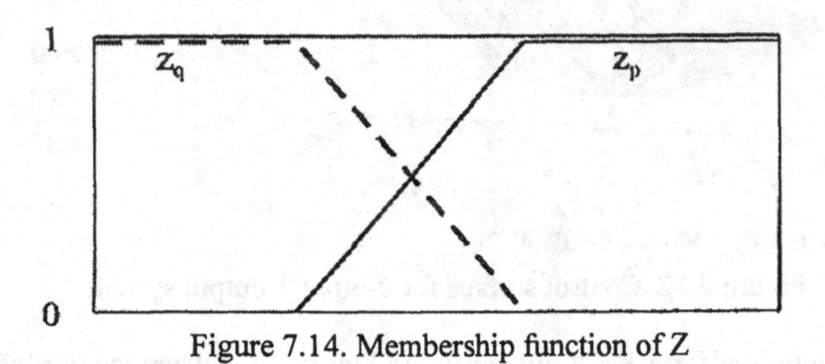

Figure 7.14. Membership function of Z

Table 7.2. Rule base for *2*-input *1*-output system

IF	IF	Then
Input of Y_a	Input Z	Output of Y_a
n_a	Z_p	n_a
ns_a	Z_p	ns_a
$zero_a$	Z_p	$zero_a$
ps_a	Z_p	ps_a
p_a	Z_p	p_a
pb_a	Z_p	pb_a
Input of Y_b		Output of Y_b
n_b	Z_q	n_b
ns_b	Z_q	ns_b
$zero_b$	Z_q	$zero_b$
ps_b	Z_q	ps_b
p_b	Z_q	p_b
pb_b	Z_q	pb_b

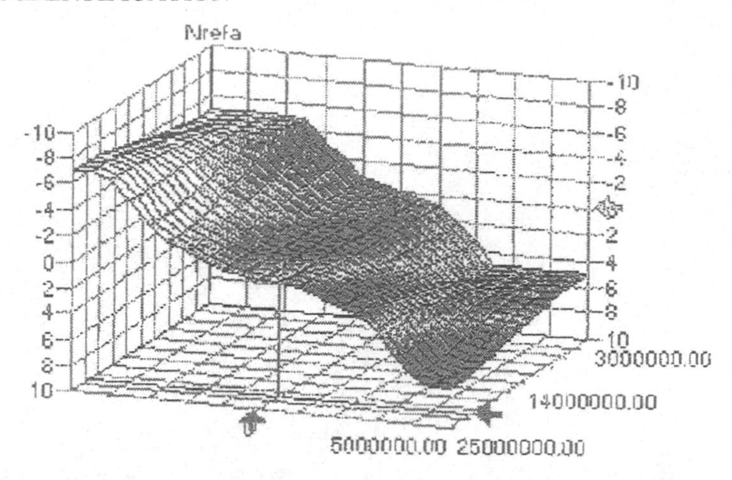

Figure 7.15. Control surface desired for *2*-input *1*-output system

References

CHILVERS RAH: "Fuzzy Control of High Power Arc Furnace. June 1996. M.Sc.Thesis, Rand Afrikaans University, Johannesburg, Republic of South Africa.
HARRIS CJ, MOORE CG, BROWN M: "Intelligent Control: Aspects of Fuzzy Logic and Neural Nets". World Scientific Publ., 1993, Singapore, ISBN 981-02-1042-6.
KOSKO B: *FuzzyThinking*. Hyperion, New York, 1933.
PROCZYK TJ, MAMDANI EH: "A Linguistic Self-Organizing Process Controller." Automatica, 1979, Vol 15, pp 15-30.
SUGENO M, KANG GT: "Structure Identification of Fuzzy Model." Fuzzy Sets and Systems, 1988, Vol 28, No 1, pp 15-33.
TAKAGI T, SUGENO M:"Fuzzy Identification of Systems and Its Applications to Modeling and Control."IEEE Trans.Sys.,Man, Cybern., 1985;SMC-15;1;116-132.,

8 STABILITY ANALYSIS OF FUZZY CONTROL SYSTEMS

8.1 Introduction

One of the prime concerns facing the designer of any new control system is related to the stability of the system after the controller is introduced. Since in most cases the controller relies on a negative feedback loop, the possibility of system instability exists whenever the loop gain exceeds unity and the loop phase shift exceeds *180* degrees. Every new design needs to be checked to ensure that instability will not occur and every existing design must contain safeguards that unstable modes will be avoided under all operational conditions. The problem of stability in closed-loop control systems has received a considerable degree of academic interest and there are a number of well-established theories to determine the stability limits of conventional controllers. However, with fuzzy controllers the situation is different. In mathematical terms, a fuzzy logic system is a mapping of an input space R^n to an output space R^m with the following properties:

- *deterministic* (the same input condition always results in the same output condition)
- *time-invariant* (the input-output function describing the mapping does not change over time)
- *nonlinear* (the output variables are not a linear combination of the input variables).

Fuzzy controllers are normally applied to practical situations where conventional control solutions either cannot be applied due to the lack of an analytical plant/process model, or where the plant/process is complex, nonlinear, and/or time-variant. In addition, the fuzzy controller itself cannot be expressed in mathematical terms, which makes a conventional analytic solution impossible. On the other hand, one must keep in mind, that an analytic stability analysis of conventional controllers *with a similar degree of complexity, nonlinearity and time-variance* is also extremely difficult or even impossible!

The question should arise: would the lack of an analytical method whereby to predict stability relegate fuzzy control to the ash heap of impractical techniques? According to many seasoned practitioners, and even some notable researchers with industrial experience, raising it as a serious issue is a typical manifestation of the gap between theoreticians and practitioners. For example, in Mamdani's opinion, the insistence on mathematical stability analysis as a condition for the acceptance of a well-designed control system exists only in academic circles. There are, in fact, no industry-approved guidelines that include stability analysis. A rigorous methodology that includes stability analysis as an essential system validation step would, at any rate, have a very limited use because few systems would be amenable to it.

So what is being done in practice to ensure stability? In acceptance tests and the commissioning of industrial control systems, *prototype testing under all plausible operating conditions takes the place of mathematical stability analysis*. If this practice has been followed in the case of conventional control systems, then according to a pragmatic view, there is no reason as to why it could not be followed for fuzzy control systems. However, to gain an extra measure of confidence, a practical well-proven and easy-to-implement method will be presented below as an aid to establish the stability limits of closed-loop fuzzy control systems. In turn, this method will serve as a diagnostic aid in the design stages to provide an insight as to what must be done to remedy any instabilities present.

8.2 Stability analysis of a fuzzy system

A closed-loop system is deemed stable if it settles at an equilibrium point after an external or internal disturbance. If the system gains are not large enough to sustain an oscillation or if the phase shift is not large enough to cause the feedback to be out of phase with the input signal, then instability cannot occur. The limiting conditions for closed-loop system stability are usually expressed mathematically in terms of the loop gain and loop phase shift as follows:

$$K_{loop} \leq 1 \text{ and } \varphi_{loop} \leq -\pi \tag{8.1}$$

System *dead time* (i.e.the delay between process input and output) is a major contributor to the loop phase shift, thus industrial plants/processes that usually have considerable dead time are more prone to instability. A qualitative explanation would be that the output is the consequence of an input which occurred sometime ago. Thus the output does not contain information about current events, only about events that happened in the past. If this output is being fed back and combined with the current input (that is happening now) to generate a control signal, then the feedback signal and the input signal will be out of phase and it may be that depending on this phase difference, instead of *negative* feedback (which is what we want because of its stabilizing effects) we may end up with a *positive* feedback which would cause the system to oscillate. (In an oscillatory state the output is no longer a function of the input but is internally generated by the system.) The fuzzy controller itself contains no memory (i.e. storage capability) and it represents a *static nonlinearity*. Thus it contributes to the system gain but not to the system delays. The fact that in general the fuzzy controller has no internal memory can be clearly seen from the so-called compositional rule of inference, whereby all fuzzy output estimates are generated from input components which are instantly available at the fuzzy estimator input.

8.3 The Vector Field Method

The Vector Field Method by Ruger et al., is applicable to the analysis of general multi-loop systems and nonlinear plants which can be *transformed into an open-loop system with only one input and one output variable*, as in Figure 8.1. It is a theoretically well-founded yet empirical method applicable to multi-input multi-output fuzzy systems.

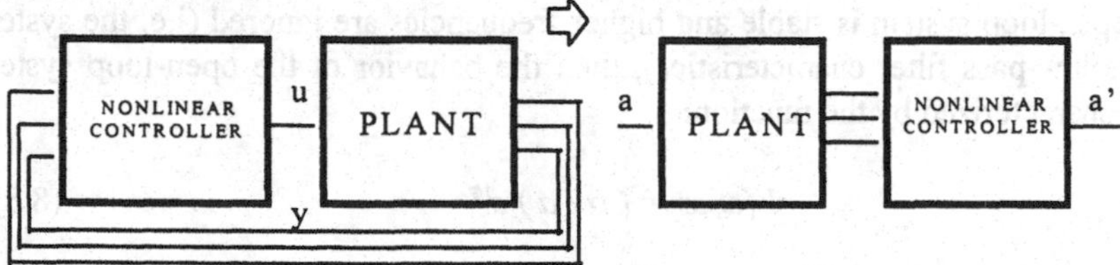

Figure 8.1. Transformation of multi-loop control system into a single-input single-output open-loop system

The nonlinear control system is assumed to be described by a symmetric output function (called an *odd* function in mathematics):

$$u = F(y) \tag{8.2}$$

and $$F(-y) = -F(y) \tag{8.3}$$

For example, consider the sine function:

$$sin\ (+x) = +\ sin(x),\ but\ sin\ (-x) = -\ sin(x) \tag{8.4}$$

whereas, for example, the cosine function is a non-symmetric (*even*) function:

$$cos\ (+x) = +\ cos\ (x)\ and\ cos\ (-x) = +\ cos\ (x) \tag{8.5}$$

The need for a symmetric output function is a constraint introduced to avoid the problems caused by the *offset* (i.e. a certain bias similar to a constant level on which the signal output is superimposed) which would be introduced by a non-symmetric function. A symmetric function has the same output response in the positive and negative directions, whereas a non-symmetric function has different responses in the positive and negative directions. The consequence is that when the open-loop system of a symmetric function is subjected to a sinusoidal input, the output varies symmetrically about the set point. For a non-symmetric function, the output will drift up or down depending on the degree of assymmetry. For small assymmetries, the difference can be ignored, but for larger ones the bias must be taken into consideration. In the following derivation, a symmetric output is assumed.

Consider the open-loop system with input a and output α, transformed from a multi-loop system where the plant/process input is u and the plant outputs are y. If this open-loop system is subjected to a harmonic input function

$$a(t) = a.e^{j\omega t} \tag{8.6}$$

then the output produced will also be a harmonic function of a different amplitude and phase shift relative to the input function:

$$\alpha(t) = \alpha\,.\,e^{j(\omega t+\Phi)} \tag{8.7}$$

If the open-loop system is stable and higher frequencies are ignored (i.e. the system exhibits low-pass filter characteristics), then the behavior of the open-loop system can be characterized by the function:

$$V(\omega,\alpha) = (\,\alpha\,/\,a\,)\,e^{j\varphi} \tag{8.8}$$

which, in general, is complex-valued. The stability limit occurs when:

$$V(\omega,\alpha) = 1 \tag{8.9}$$

Pairs of values (ω,α) for which Equation 8.9 is fulfilled, indicate that a *limit cycle* of amplitude a and frequency ω probably exists in the closed-loop system. The system is stable for all cases when

$$V(\omega,\alpha) < 1 \tag{8.10}$$

8.4 Analysis procedure

In practice, the values of the function $V(\omega,\alpha)$ usually cannot be calculated analytically, but can be determined by simulation or experiment for discrete pairs of values (ω_l, α_j) which are grid points in the $(\omega\text{-}\alpha)$ plane. Based on these values, using an interpolation method the function values of $V(\omega,\alpha)$ can be determined for any point (ω,α). In order to find points (ω,α) that satisfy Equation 8.9, i.e. the point where a limit cycle occurs, the complex-valued function $V(\omega,\alpha)$ is interpreted as a *vector field* and visualized in the (ω,α) plane. For this, we calculate and plot the *contour curves* of the functions :

- $abs\,[V(\omega,\alpha)] - 1$ *(contours of constant gain)*
- $arc\{\, V(\omega,\alpha)\,\}$ *(contours of constant phase shift)*

A certain point (ω,α) for which

$abs\,[V(\omega,\alpha) - 1] = 0$ *(system gain = 1)*

$arc\{\, V(\omega,\alpha)\,\} = 0$ *(phase shift = 180^0)*

these curves *intersect*, meets the condition of Equation 8.9, that is, a *limit cycle occurs*. In an actual case, the system simulator is run with various different sets of input values using different frequencies and amplitudes to obtain a set of points for the function $V(\omega,\alpha)$. For each, the corresponding output amplitude and phase shift are to be measured, tabulated and plotted. Figure 8.2 shows an example for the occurrence of a limit cycle with intersecting curves, while Figure 8.3 illustrates a case with no intersecting curves, i.e. no limit cycle.

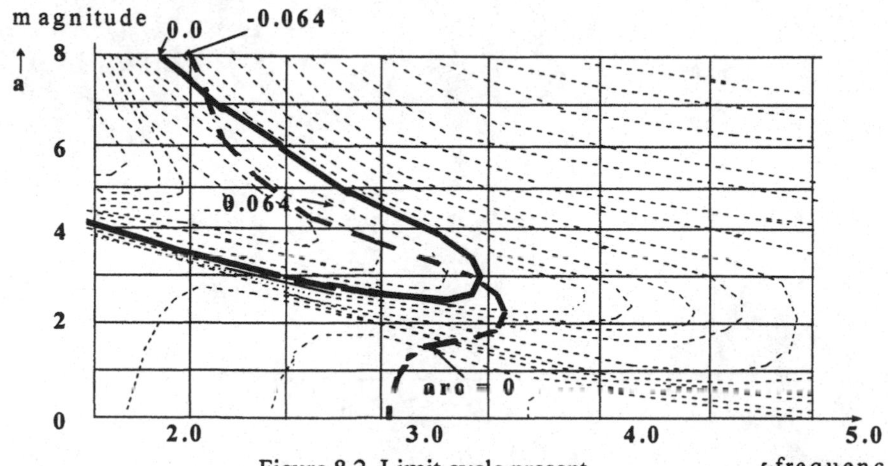

Figure 8.2. Limit cycle present

(Reprinted with the permission of the ELITE Foundation, Aachen, Germany)

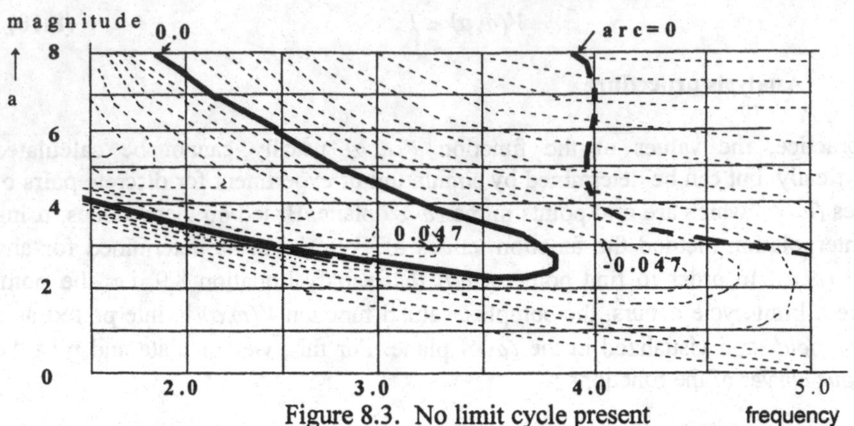

Figure 8.3. No limit cycle present

(Reprinted with the permission of the ELITE Foundation, Aachen, Germany)

References

MALKI HA, LI H,CHEN H:"New Design and Stability Aznalysis of Fuzzy Proportional-Derivative Control Systems."IEEE Trams.on Fuzzy Syst.,1994;2;4;245-254.

MAMDANI EH: "Twenty Years of Fuzzy Control: Experiences Gained and Lessons Learnt."

VON ALTROCK C: *Fuzzy Logic and Neurofuzzy Applications Explained.* Prentice Hall, 1995, Englewood Cliffs, NJ, USA

OPITZ HP: "Fuzzy Control and Stability Criteria." First European Congress on Fuzzy and Intelligent Technologies, EUFIT'93), Sep 1993, Aachen, Germany.

RUGER JJ, FRENCK CH, MICHALSKE A, KIENDL H: "The Vector Field Method for Stability Analysis: A Case Study." Second European Congress On Fuzzy And Intelligent Technologies, EUFIT'94 Conference Sep 1994, Aachen, Germany pp 291-294.

SINHA NK:"Control Systems." Wiley Eastern Ltd. 1994.London.

STEPHANOPOULOS G: *Chemical Process Control*. Prentice Hall Internat'l Edition,1984.

9 NEUROFUZZY CONTROLLERS

9.1 Why neurofuzzy?

Section 5.2 listed three fuzzy system types. The first was the rule-based system where human operator experience is incorporated into a rule set and membership functions. The second (parametric) and third (relational equation based) have reduced the element of human judgment to the construction of membership functions while the rules were generated from measurements.

A neurofuzzy system uses a neural network to generate both the rules and membership functions from measurements. In this way, in principle it would be possible to do away with human experience altogether. However, a neurofuzzy software development system (such as the generic system described in Chapter 10), needs to begin the design process with an initial fuzzy system and at this stage it is still necessary to incorporate human knowledge, if available, into this initial structure. Some qualitative *a priori* knowledge about an industrial system always exists and should therefore be incorporated in designing a neurofuzzy controller. A neurofuzzy controller software development system should make provisions for

excluding from neurofuzzy development those rules that resulted from human judgments. This combination of quantitative and qualitative knowledge is the chief advantage of neurofuzzy control systems.
Another way of using a neurofuzzy design is to doublecheck the accuracy of the rules and membership functions based on human operator experience. After a short introduction to elementary neural network theory, this chapter aims to provide a theoretical and practical understanding of the neurofuzzy structure, and of the design procedure using it. The following will provide only a basic treatment of neural networks sufficient for the understanding of neurofuzzy control systems.

9.2 Neurons and neural networks

Neural networks (or better, *artificial* neural networks) resemble their biological counterparts. In contrast to traditional computing machines which rely on one or more relatively complex processors, neural networks consist of a large number of simple computational units, so-called *neurons* (Figure 9.1), highly interconnected with communication pathways.

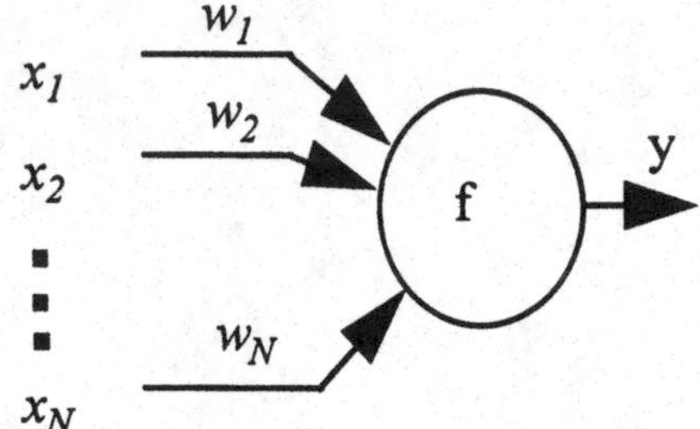

Figure 9.1 Model of a typical artificial neuron

The output of this neuron is a nonlinear function of the weighted sum of its inputs. The input-output relationship is defined by

$$y = f\left(\sum_{i=1}^{N} w_i \, x_i - \theta \right) \tag{9.1}$$

where the w_i 's are the weights, θ is a threshold and f is some nonlinear function, the *activation function*, such as for example, those shown in Figure 9.2 below.

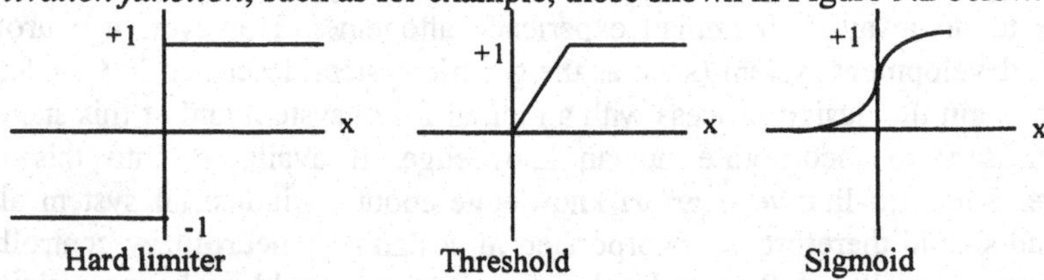

Figure 9.2. Typical input-output functions of neurons.

The most frequently used topology in neurofuzzy controllers is the *feedforward* type whereby each output in a layer is connected to each input of the next layer. This arrangement may be expressed as follows:

Output layer: $$y_k = \sum_{j=1}^{M} w_{ki}\, v_i \tag{9.2}$$

Hidden layer $$v_i = f\big(\sum_{j=1}^{M} \alpha_{ij}\, z_j\big) \tag{9.3}$$

Next hidden layer if any: $$z_j = f\big(\sum_{i=1}^{N} \beta_{jl}\, u_l\big) \tag{9.4}$$

where the sigmoid input-output function $f(x) = 1/\ (1 + e^{-ax})$ is assumed and the locations of variables *y, v* and *z* are shown in Figure 9.3. With a sigmoidal function, the output limits are *0* and *1*.Each neuron in a neural net processes the incoming inputs that enter via the *synapses* (i.e. input "terminals") and sends them to an output which, in turn, is connected to another neuron. The neural net shown in Figure 9.3 has a *layer* for *input* signals, two *hidden* layers and one layer for *output* signals. The information enters at the input layer, (i.e. connections from the inputs), is processed by the internal layers and emerges at the output layer. The objective of a neural net is to process information according to its previous *training with example input-output data.* This training actually forms the neural net structure prior to using it for computation.

Neural networks are *highly parallel*; their numerous operations are executed simultaneously. Parallel systems can execute hundreds or thousands of times faster than conventional microprocessors that process data sequentially, thereby making many embedded neural network applications practical.

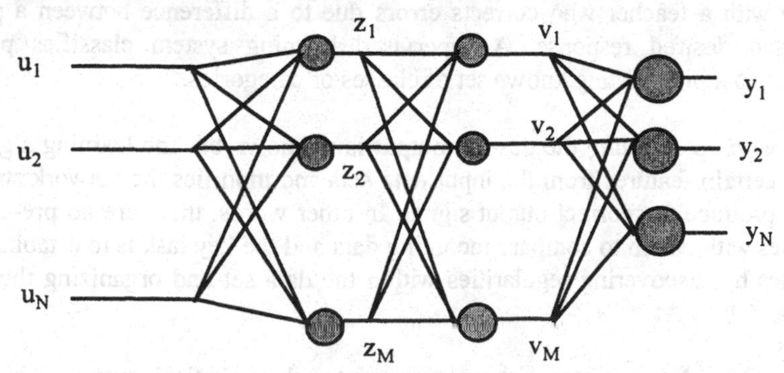

Figure 9.3. Feedforward neural network

For the sake of simplicity, in Figure 9.3 it is assumed that each neuron in the hidden layers has the same sigmoidal input-output function. However, this is not a necessary requirement. Since there is no feedback between layers, the feedforward topology produces a nonlinear mapping from input to output nodes. Each interconnection has a *weight* associated with it, such as w, α and β. Such a neural network has two operational modes: *training* (or learning) and *estimating*.

9.3 Training phase

In the training phase, the objective is to set up the network structure according to some desired property by adjusting the interconnection weights. In turn, the network is used to estimate the output from its inputs on the basis of the thus formed structure. Training uses sample data sets of *randomized* inputs and corresponding outputs. For this the net uses *learning algorithms*.

Learning means encoding a pattern in the learning system's structure. That is, the system structure changes whenever the system learns new information. Learning is different from data acquisition where the new information patterns are merely stored for further processing. For example, in the self-learning relational equation based fuzzy system described in Chapter 6, the relation matrix is entirely recalculated upon the arrival of every new data item, thus incorporating this new information into the fuzzy system's structure (i.e. knowledge base). We have seen that in a neuron model the strength of an input depends on the weight of its neural pathway (synapse). Repeated activation of this path increases while a decreased level of activation lowers the synaptic weight.[1]

There are several different kinds of learning algorithms, as has been briefly indicated in Section 7.2, such as supervised, unsupervised, competitive, differential competitive, etc. each of which has certain merits and drawbacks. Lack of space does not permit an exhaustive treatment of these algorithms. However, a qualitative description of the first two will be restated here, as follows.

In *supervised learning*, the desired neuron outputs are known. The difference between the actual and the desired output constitutes an error signal which is used to update the weights according to the specific algorithm involved. This is analogous to learning with a teacher who corrects errors due to a difference between a pupil's actual and desired response. A supervised learning system classifies patterns according to a pre-existing known set of classes or categories.

In *unsupervised learning* the desired output is unknown but the learning algorithm extracts certain features from the input data sets and modifies the network structure so as to produce the correct output signal. In other words, there are no pre-existing categories with which to compare incoming data and the key task is to establish such categories by discovering regularities within the data set and organizing them into clusters.

Example 9.1: Assume we wish to construct a hypothetical system which can recognize whether English or German is being spoken. There is a lightbulb that lights up for English and another one for German. Initially the wrong lightbulb will light up in most cases but by judicious tuning of the system (analogous to the action of the learning algorithm) the correct light bulbs will burn whenever the corresponding language is spoken. The system extracts certain grammatical features of the spoken sentences like, for example, the fact that in German the verb usually (and recurrently) occurs at the end of a sentence.

Initially the net is inert and the *learning algorithm modifies the individual weights of their interconnecting pathways* in such a way that the network behavior reflects the desired one. Thus the network can incrementally alter its structure in time until it achieves the desired performance. The *training phase* of a neural network is analogous to the learning phase of a self-learning fuzzy system discussed in Chapter 6. As a result of training, the neural network will output values similar to those in the sample data sets when the input values match the training samples. For values in between, it interpolates. In other words, it can *learn from examples* presented to it!

9.4 Estimating phase

The training phase is followed by the *estimating phase* where the neural network behavior is *deterministic*, i.e. for every combination of input values the output value is always the same. The trained network can estimate, recognize and classify unknown or incompletely known input data patterns. They infer solutions from the data presented to them, often discovering and capturing subtle relationships. This ability differs radically from standard software design techniques because *it does not depend on a programmer's prior knowledge of all possible answers*.

9.5 Pattern recognition in the presence of noise

While conventional computers are good in performing complex calculations accurately and handle data sequentially, neural networks are good at shape recognition, speech synthesis, cleanup of noisy signals, perceiving partially hidden objects in a complicated scene, or recognizing subtle relationships which are not obvious to human observers. In addition, neural networks can learn underlying relationships even if they are difficult to find and describe. Neural networks can *generalize*. This means that they can extract a useful output from incomplete, imperfect or noisy data. This feature is very useful in practical applications where data is noisy. They can capture complex, higher-order functions and nonlinear relationships among input variables. Their noise tolerance is attributable to the averaging effect of the parallel processing of input data.

9.6 Fault tolerance

Inherent *fault tolerance* is another advantage. In a conventional computer if one part fails, usually the whole system breaks down, while in a neural network fault tolerance is built in, due to a distributed processing structure. If one neural element fails, its erroneous output is "swamped" by the correct outputs of its neighboring elements.

9.7 Backpropagation algorithm

There are many different types of neural networks, all of which perform the same fundamental function of mapping an input vector to an output vector of one or more components, each of which represents the value of some variable. Input-to-output mapping may be *memoryless* (i.e. there are no storage elements), in which case a *feedforward* network will suffice. Or it may be *dynamic*, involving *previous states* (i.e values stored in a memory), when a feedforward network with delayed inputs or a feedback network is used. In recent years, the feedforward backpropagation algorithm has gained dominance, especially in neurofuzzy control applications. It is therefore important to understand its internal mechanisms, both qualitatively and analytically.

9.7.1 A qualitative description of backpropagation

The backpropagation algorithm is a *learning algorithm*. As has been shown earlier, a *feedforward* topology, using the backpropagation algorithm *for training*, has one *input* layer, at least one *hidden* layer, and one *output* layer. A *backpropagation* network operates in two steps *during training*. First, an input pattern is presented to the network's input layer. The resulting activity flows through the network from layer to layer until the output is generated. Next, the network output is compared to the desired output for that particular input pattern. The error is passed backwards through the network — from the output layer back to the input layer, with the weights being modified as the error propagates backwards. This signal passes through the *activation function* which is typically smooth and nonlinear, like the sigmoidal function given earlier and depicted in Figure 9.3.

The backpropagation algorithm, used for training the neural network, aims to first *estimate the error*, i.e. the difference between the actual and the desired values of the output for a given input stimulus and, in turn, *reduce this error* by changing the weights attributed to each pathway between neurons residing in various layers.

The first task is to carry out an approximation describing the least-square convergence (i.e gradient) of an *estimator* to a minimum of an error function, which represents the error at iteration k at the network output. Since the desired output is known during training, we speak of *supervised learning*. This approximation estimates the gradient at each iteration in a discrete *gradient-descent algorithm*. The gradient points in the direction of steepest descent on the expected-error surface. When the gradient is zero, a minimum has been reached. Unfortunately there is no guarantee that this minimum is the *global minimum*; it might only be a *local minimum* even though this is said to happen rarely. A local minimum is unfavorable because in this state the network can accept no more training, even if such training would still be needed for further performance improvement, thus resulting in suboptimal learning.

Let us look at the internal mechanism of the backpropagation algorithm. The learning algorithm for the backpropagation network is the *Generalized Delta Rule* which is an application of the least-mean-square (*LMS*) method. The name *Delta* alludes to the difference between a present and a previous value while the smallest mean square difference between them is to be searched for. Here the error is used not just to affect one set of weights, i.e. input-to-output, but two sets: input layer-to-hidden layer and hidden layer-to-output layer. The *Generalized Delta Rule* takes advantage of the chain rule of differential calculus to compute the way in which these weights depend on one other. This is accomplished in two stages as follows.

In the first stage, the weights between the hidden and the *output* layers are adjusted. This stage is, in fact, driven by the difference between the actual and desirable values at the output nodes.

In the second stage, the weights between the input and the *hidden* layers are adjusted. This is different from the previous adjustment, because the designer does not know what the desired values of the hidden nodes should be.

9.7.2 Analytical description of backpropagation

The goal of this weight adjustment process is to reduce the error E at the output, defined as the *LMS* error. The basic idea is to change the weights in the direction of decreasing energy, where energy means some measure of the errors at the output. Assume that the system's output is y_k^d where the subscript denotes the kth element of the desired output vector y_d. The error is then $\delta y_k = (y_k^d - y_k)$ and the energy we wish to minimize is the mean square error:

$$E = (1/2) \sum_{k=1}^{N} (\delta y_k)^2 = (1/2) \sum_{k=1}^{N} (y_k^d - y_k)^2 \tag{9.5}$$

Since we wish to minimize the error, we can use differential calculus to obtain the minimum, provided that the error equation operates on *differentiable* functions[2] Let us express the training rule in terms of a function Δ which is applied to every connection weight w_{ij} after each presentation of the training pattern. .A repeated application of this function will hopefully make the network to converge at a set of weights which minimizes the error for recognizing all patterns of the training set. The function Δ is related to both the error and the weights as follows:

$$w_{ij,new} = w_{ij,old} + \alpha\, \Delta(w_{ij,old}) - output_{act\,j} \tag{9.6}$$

where: w_{ij} = new and old values of connection weight between nodes$i$ and j;

α = constant

$\Delta(w_{ij,old})$ = a function value, calculated for each $w_{ij,old}$ which determines the amount of change to be made in each weight;

$output_{act,j}$ = output activation level of the jth neuron.

Let us set the Δ function to be proportional to the negative error derivative with respect to the connection weight:

$$\Delta(w_{ij}) = -\beta\, \partial E / \partial w_{ij} \tag{9.7}$$

where w_{ij} represents the weight w_i of input x_j ; $\Delta(w_{ij})$ is the incremental change in a particular weight, β is a learning rate coefficient and E is the error function with respect to the weight which is being modified.

Consider the relationship between the total error, E, and the weights w_{ij} . The value of E is calculated as a discrete value after having presented a training pattern to the network. Assume that there exists a value w_{ijmin} , for which E is minimum and that our task is to find it. The derivative of E with respect to w_{ij} *is* the tangent of the error curve at any point defined by w_{ij} . Figure 9.4 shows that if the error monotonically decreases as w_{ij} approaches w_{ijmin} , then using this method (called *grandient descent* method) incrementally can find w_{ijmin} . The steeper the curve the more rapidly the minimum will be reached, while if the curve is shallow, convergence will be slow. However, a problem can arise if the error does not monotonically decrease like in the right-hand side of Figure 9.4 where a shallower minimum at $w_{ijmin}*$ is shown. If the gradient method is applied in this vicinity, then the method will converge at $w_{ijmin}*$, *a local minimum,* and not at w_{ijmin} the *global minimum*. If the system converges to the local minimum, it "gets stuck" as mentioned before. There are several practical techniques to get out of this predicament which fortunately occurs very rarely[3]. Commercial software based neural nets use special algorithms to circumvent this problem. As for neural net and.or neurofuzzy software written "in-house", the following would be helful.

How does one notice that the neural network portion that generates membership functions and rules for the fuzzy system is performing suboptimally? One of the most effective methods is to repeat neural training with different initial conditions. This is because if the starting point of neural training defined by the initial conditions is near a local minimum, a gradient descent based training tends to find this one as the solution. In as much as the neural network modifies the membership functions and the *Degree-of-Support* of the rules of a pre-existing fuzzy system, one can always effect manual modifications of these factors. Starting from a different point by adding some rules or using randomly distributed *Degrees-of-Support* of the rules in the initial fuzzy system may clear up the problem. Selecting a *random learning rate* from among the learning algorithms provided by the software development system used can also help. In fact, neural training and manual changes can be achieved interactively. Thus in practice local minima certainly do not represent an unsolvable problem, for they only make the design procedure somewhat more cumbersome. We shall now present an analytic derivation of the principle of backpropagation .

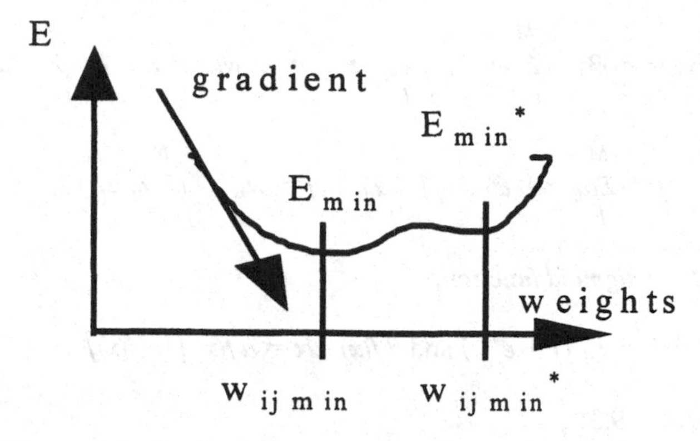

Figure 9.4. Gradient descent with global and local minima.

The following derivation shall apply to the feedforward neural network with one hidden layer. According to Equations 9.5 and 9.7, the weight changes should be chosen as:

$$\Delta w_{ki} \propto - \partial E / \partial w_{ki} ; \; \Delta \alpha_{ij} \propto - \partial E / \partial \alpha_{ij} ; \tag{9.8}$$

Output layer weights
Take the first term:

$$\partial E / \partial w_{ki} = \partial / \partial w_{ki} \left[(1/2) \sum_{k=1}^{N} (y_k^d - y_k)^2 \right] = - \sum_{k=1}^{N} \delta y_k \; \partial y_k / \partial w_{ki} =$$

$$= - \sum_{k=1}^{N} \delta y_k \; (\partial / \partial w_{ki}) \left[\sum_{k=1}^{M} w_{ki} \, v_i \right] = - \; \delta y_k \, v_i \tag{9.9}$$

thus the update value of the weight w_{ki} is $\Delta w_{ki} = \eta_1 \; \delta y_k \; v_i$, where η_1 is a proportionality constant . The new weight of the connection (k,i) is then

$$w_{kinew} = w_{kiold} + \Delta w_{ki} \tag{9.10}$$

while $\quad \delta y_k = y_k^d - y_k = e_k =$ the error at the output $\quad$ (9.11)

Hidden layer weights
Consider the weights α_{ij} and compute the negative gradient of the error:

$$\partial E / \partial \alpha_{ij} = - \sum_{k=1}^{N} \delta y_k \;\; \partial y_k / \partial \alpha_{ij} \tag{9.12}$$

First evaluate the second term on the right:

$$\partial y_k / \partial \alpha_{ij} = \partial / \partial \alpha_{ij} \left[\sum_{i=1}^{M} w_{ki} \; f\left(\sum_{j=1}^{M} \alpha_{ij} \; z_j\right)\right] = \sum_{i=1}^{M} w_{ki} \; \partial / \partial \alpha_{ij} \left[f\left(\sum_{j=1}^{M} \alpha_{ij} \; z_j\right)\right] =$$

$$= \sum_{i=1}^{M} w_{ki} \; f' \left(\sum_{j=1}^{M} \alpha_{ij} \; z_j\right) \partial / \partial \alpha_{ij} \left[\sum_{j=1}^{M} \alpha_{ij} \; z_j\right] = w_{ki} \; f' \left(\sum_{j=1}^{M} \alpha_{ij} \; z_j\right) z_j \tag{9.13}$$

Assume that f is a *sigmoid function*:

$$f(x) = 1 / (1 + e^{-ax}) \text{ and } d f(x) / dx = a f(x) [1 - f(x)] \tag{9.14}$$

and from Equation 9.3:

$$v_i = f\left(\sum_{j=1}^{M} \alpha_{ij} \; z_j\right) \tag{9.15}$$

Then from Equations 9.14 and 9.15:

$$f' \left(\sum_{j=1}^{M} \alpha_{ij} \; z_j\right) = a \, v_i \, (1 - v_i) \tag{9.16}$$

which yields

$$\partial y_k / \partial \alpha_{ij} = w_{ki} \; [a v_i \, (1 - v_i)] z_j \tag{9.17}$$

and substituting Equation 9.17 into 9.12:

$$\partial E / \partial \alpha_{ij} = - \sum_{k=1}^{N} \delta y_k \; [a v_i \; (1 - v_i] z_j = - a \, v_i \; (1 - v_i) \left[\sum_{k=1}^{N} w_{ki} \; \delta y_k\right] z_j \tag{9.18}$$

Finally, define δv_i as the backpropagated error at the hidden layer:

$$\delta v_i = a v_i \, (1 - v_i) \sum_{k=1}^{N} w_{ki} \; \delta y_k \; , \; k = 1,2,\ldots,N; \; i,j = 1,2,\ldots,M; \tag{9.19}$$

then the update value of the weight α_{ij} *is*

$$\Delta \alpha_{ij} = \eta_2 \; \delta v_i \; z_j \tag{9.20}$$

where η_2 = proportionality constant. The new weight of the connection (i, j) is:

$$\alpha_{ijnew} = \alpha_{ijold} + \Delta \alpha_{ij} \tag{9.21}$$

In practice, these steps of the algorithm are repeated until the average error becomes less than a preset threshold, or the number of preset iterations have been completed. Other parameters such as the *learning rate coefficient* are usually also adjustable. There are a number of neural network software development systems available and attempts have also been made to construct hardware chips for neural networks. To date, such systems have not yet been widely used in industry due to various shortcomings, to be elaborated on in the next section.

9.8 Neural networks: strong and weak points

Neural networks offer intelligent features such as learning, adaptation, fault-tolerance, and generalization. Since their main strength lies in pattern recognition, their current principal applications are in optical character recognition, function estimation, financial forecasting and, at least potentially, in process control. Interestingly, neural networks can be trained to become predictive process models. Many industrial firms routinely collect process data by logging the outputs of various sensors and gauges. In the absence of mathematical models, neural networks can utilize the thus collected historical data to build non-mathematical process models. After training with such data, a neural network can become the model which can predict process reactions. However, neural networks, if used alone, (i.e not combined with fuzzy systems) have certain shortcomings which have had a strong dampening effect upon their widespread application. These factors are described here as follows.

- It is not possible to trace the manner in which a neural network has arrived at a particular result. In this sense, neural networks are like human experts who express opinions often without being able to easily explain them. In other words, you cannot look into someone's head to see the process of thinking as to how a particular result was obtained; there is no "audit trail" to follow and there is no easy way to verify and optimize an output.
- Since one cannot look into the neural network, it is impossible to interpret the causes of a particular behavior . One can also not modify the network manually to change to a more desirable behavior. This inflexibility is one of the prime reasons for the relative lack of industrial applications.
- Training methods are imperfectly understood and besides trial-and-error methods, there are few guidelines for designers to follow. For example, defining the number of hidden nodes, or establishing convergence to a given tolerance range depends on experimentation. In an industrial setting there is little room for such experimentation.
- Training time is unpredictable and it may be inordinately long. For this reason, on-line retraining (i.e updating) usually required for a real-time self-tuning system is not a good application for neural networks.
- Execution times (i.e. the estimating phase) depend on the number of connections and is roughly equal to the square of the number of nodes used. This means that just a few more nodes can considerably increase the execution time. However, specialized neural network hardware has recently become available to alleviate this problem.
- Neural network chips are not readily adaptable to the mass market because the large number of calculations would result in prohibitively large chip size.
- We discussed in detail only the backpropagation model because of its popularity in neurofuzzy control. However, in general the neural network designer may have to choose from among many different learning algorithms suitable to the application involved. There is unfortunately no clear-cut method for choosing the right model except for trial-

and-error. Some neural network software packages offer as many as *30* different network topologies and learning algorithms as an aid to designers.

- The theoretical aspects of problems associated with local versus global minima have been solved in commercial neural network packages by special techniques (see Note 2). In the case of self-written neural network software, the problem can be overcome by changing the initial training conditions.

9.9 Fuzzy logic and neural networks as intelligent tools

Fuzzy logic theory provides the foundation for capturing the uncertainties associated with human thinking processes by employing linguistic definitions of variables used in a rule-based inference engine. It can describe the desired control behavior with simple **IF...THEN** relationships, usually from interviews with human operators. Fuzzy control systems can be easily tested, verified and their specific behavior can be changed and optimized. However, the main drawback of a fuzzy system is that all aspects of control behavior must be defined explicitly (i.e. converted into rules and membership functions) by means of human operator interviews and from existing *a priori* process knowledge. In the absence of experienced human operators , the desired behavior is often contained in data sets and the designer has to extract the **IF...THEN** rules from the data manually by means of a difficult and time-consuming process. On the other hand, neural networks, trained with the data sets that contain the desired system behavior can extract such information and pass it on to the fuzzy controller. As shown in Table 9.1, *the combination of the two technologies provides a complete answer to intelligent controller design problems.* This is the reason for devoting this book to fuzzy and neurofuzzy solutions and affording only a perfunctory look at the use of neural networks on their own.

According to convention, neurofuzzy systems consist of a neural network based "front end" which generates rules and/or membership functions for the fuzzy system that uses this information. The generic fuzzy controller software development system described in Chapter 10 has a neural network option whereby to facilitate the design of a neurofuzzy system in a systematic manner. In essence, the designation *neurofuzzy* has gained popularity in industrial systems and it means the *incorporation of a neural network into an existing fuzzy system.*

9.10 Structured neurofuzzy design

In a neurofuzzy control system, it is necessary to generate an initial fuzzy controller with membership functions and rules that reflect the designer's *a priori* knowledge of the plant/process. In turn, the neural network generates modifications of the initial membership functions and fuzzy rules and these are passed on to the fuzzy controller. A more detailed explanation of such modifications will be given in Chapter 10. Figure 9.5 shows a block diagram depicting the systematic training of neural networks in neurofuzzy systems. As has been mentioned before, any knowledge of the plant/process that has an influence upon controller design should be incorporated into the initial fuzzy system. It also makes sense to group the rules that are set up for steady-state, transient, and alarm conditions in order to facilitate testing and maintenance[4]. The alarm conditions, represented by rules that affect system safety and integrity become active rarely, yet they act as watchdogs. The small blocks represent rules and membership functions related to the specific conditions denoted by the inscriptions. Provisions must also exist to exclude all of the rules discussed so far from neurofuzzy training so that the rules and membership functions that pertain to *a priori* process knowledge would not be modified by the neural network. However, even the modifiable blocks should be grouped and trained according to the specific conditions inscribed.

9.11 A qualitative summary of neural and neurofuzzy systems

- This section has reviewed the concepts of error backpropagation training and some details of the processing techniques required.

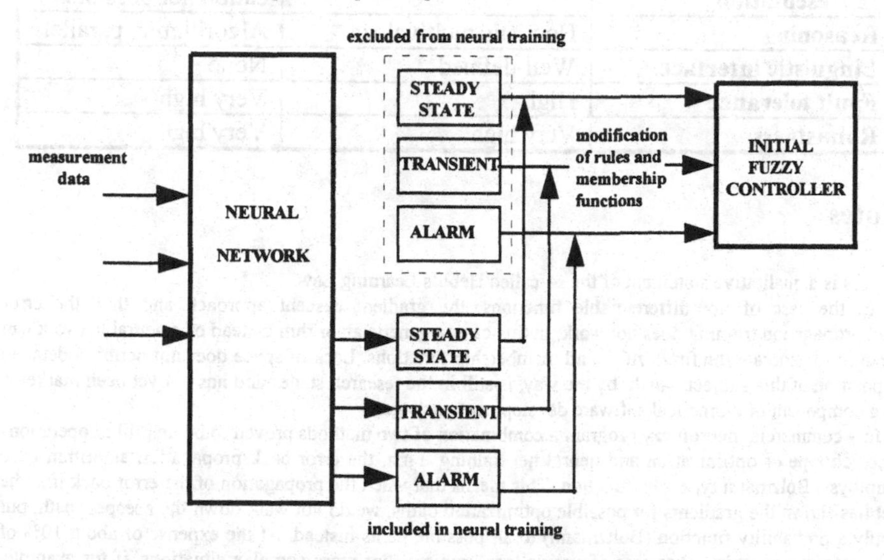

Figure 9.5. Neural training of an initial fuzzy controller.

- Fuzzy logic provides a convenient and user-friendly means to develop programs for neurofuzzy controllers, helping designers to concentrate on the functional objectives instead of mathematical details. On the other hand, artificial neural networks are suitable for intensive numerical data handling
- Fuzzy and neural technologies (often called *soft computing*) show enormous potential for applications that combine qualitative knowledge with robustness and learning.
- The neurofuzzy approach makes it possible to extract rules out of numerical data. In industrial applications, neurofuzzy techniques are very promising.

9.12 A last word on learning and estimating

From the foregoing discussions on the learning and estimating phases of all intelligent (i.e.fuzzy, neural, neurofuzzy) control systems, an important insight is that the learning or training phase is used to establish the system's knowledge base, while in the estimation phase the thus constructed knowledge base is used to estimate the system response. This is true even about rule-based fuzzy systems, because in this case learning is accomplished by the human operator who develops his own "knowledge base" to control the system. In fact, learning is actually the same as systems identification: both aim to establish the structure of a model used subsequently for estimation.

Table 9.1 Comparison between non-adaptive fuzzy and neural systems

	Fuzzy Systems	**Neural Networks**
Knowledge acquisition	Human experts	Numerical data
Training method	Precise definitions needed	Learning by example
Knowledge representation	Easily verifiable	Verification/modification not possible
Reasoning	Heuristic,multivalued	Algorithmic, parallel
Linguistic interface	Well-defined	None
Fault tolerance	High	Very high
Robustness	Very high	Very high

Notes

1. This is a qualitative statement of the so-called Hebb's Learning Law.
2. In the case of non-differentiable functions, the gradient descent approach and thus the error backpropagation training does not work. In this case, a genetic algorithm instead of a neural network can be used to generate the fuzzy rules and membership functions. Lack of space does not permit a detailed exposition of this subject which, by the way, is still in the research stages and has not yet been marketed as a component of a practical software development package.
3. In a commercial neurofuzzy program a combination of two methods proved to be helpful in operations research type of optimization and neural net training. First, the error back propagation algorithm used employs a Bolzmann type goal function. This means that when the propagation of the error back into the net has shown the gradients for possible optimization paths, we do not walk down the speepest path, but apply a probability function (Boltzmann) to all possible paths instead. At the expense of about 10% of the performance, this takes care of simple local minima. For more complex situations, if for example, the neurofuzzy module gets stuck in a whole sequence of local minima, there is a secondary mechanism called Tabu Search. This technology implements both a short-term and long-term memory for the interations. Thus, after detecting that one is in a loop one can exclude (tabu) certain paths from the optimization route. Having pulled these two technologies together, the local minima problems have been solved.
4. In principle, such a systematic grouping of rules could also be undertaken in ordinary fuzzy controller design whenever the number of rules and/or rule blocks is relatively large.

References

The following references represent a relatively small selection of the hundreds of publications that exist on neural networks. It is, however, sufficient to provide an understanding of the field as a background to neurofuzzy systems.

BOSE BK:"Expert System, Fuzzy Logic, and Neural Network Applications in Power Electronics and Motion Control." Proc.IEEE,;1994;82;8;1303-1325.

CARPENTER GA, GROSSBERG S, MARKUZON N, REYNOLDS JH, ROSEN DB, "Fuzzy ARTMAP: a Neural Network Architecture For Incremental Supervised Learning of Analog Multidimensional Maps," IEEE Trans. Neural Networks, 1992; 3;. 698-713.
COX, E: "Adaptive Fuzzy Systems". IEEE Spectrum February 1993; 27-31.
FRIEDEL P: "Fuzzy And Neurofuzzy Application in The Consumer Market - a Philips Perspective". Dataweek, January 1996.
FROESE T: "Applying Fuzzy Logic and Neural Networks To Modern Process Control Systems". SA Instrumentation and Control, September 1994.
GORI M, TESI A:"On the Problem Of Local Minima in Backpropagation.". IEEE Trans.on Pattern Analysis and Mach.Intell.1992;14;1;76-85.
fuzzyTECH 4.0 *User Manual*, Revision 402, 1995 June, Inform Corp.Aachen, Germany.
GUPTA MM, KNOPF GK: "Fuzzy Neural Network Approach to Control Systems," Proc. of First Int. Symp. on Uncertainty Modeling and Analysis," Maryland, Dec. 1990;483-488.
HALPIN SM, BURCH RF:"Applicability of Neural Networks to Industrial and Commercial Power Systems: A Tutorial Overview." IEEE Trans.Industry Appl. 1997;33;5;1355-1361.
HARRIS CJ, MOORE CG, BROWN M: "Intelligent Control: Aspects of Fuzzy Logic and Neural Nets". World Scientific Publ., 1993, Singapore.
HECHT-NIELSEN R:"Theory of the Backpropagation Neural Network." Proc.Int'l Joint Conf.Neural Networks,1989;1;593-605.
HEBB DO*: The Organization Of Behavior.*1949. Wiley, New York.
HOHLFELD M: "The Roles Of Neural Networks and Fuzzy Logic In Process Optimization". SA Instrumentation and Control, February 1994.
HOPFIELD JJ: "Neural Networks and Physical Systems With Emergent Collective Computational Abilities," Proc. Nat. Acad. Sci. U.S.A; 79; 2554-2558.
HORIKAWA S, FURUHASHI T, OKUMA S, UCHIKAWA Y: "Composition Methods Of Fuzzy Neural Networks," Conf.Rec. IEEE/IECON 1990;. 1253-1258.
JANG JSR:,SUN, CT: "Neuro-Fuzzy Modeling and Control."Proc.IEEE, 1995;83;3;378-406.
KAYA A: "An Intelligent *PID* Control Algorithm By Neural Networks". Proc. IFAC Symp.Int.Comp.and Instr.for Contr.Appl.,Malaga, Spain, May; 417-420.
KLIMASAUSKAS CC:"Applying Neural Networks, Part III:Training a Neural Network." PC-AI Magazine, May/June 1991;0-24.
KOSKO B: *Neural Networks and Fuzzy Systems*. Prentice Hall, 1992
MARCH A, CRAIG H, R PAP: *Handbook of Neural Computation*. Academic Press, 1990.
MATHER AJ, SHAW IS:" Alternative Method For the Control Of Balancing Tank At A Wastewater Plant." Water Sci.and Techn., 1993;28;11-12;523-530.
MINSKY M, PAPERT S: *Perceptrons*. MIT Press, Cambridge, Mass., 1969..
MOORE KL:*Iterative Learning Control for Deterministic Systems*.Springer, 1993.
MOORE KL:*Tutorial on Neural Networks*. IEEE Potentials, 1992;11;1;23-28.
NGUYEN DH, WIDROW B: "Neural Networks for Self-Learning Systems." IEEE Control Systems Magazine, April 1990; 18-23
PASSINO K:"Intelligent Control for Autonomous Systems." IEEE Spectrum,1995 June;55-61.
ROSENBLAT F: *Principles of Neurodynamics*. 1962 ,Spartan, New York.
RUMELHART DE, HINTON GF, WILLIAMS RJ: "Learning Internal Representations By Error Propagation," in: *Parallel Distributed Processing: Exploration in the Microstructure of Cognition, Vol. 1 Foundations*, PDP Research Group, MIT Press/Bradford Books, Cambridge, Mass., 1986.
SAERENS M, SOQUET A:"Neural Controller Based On Back-Propagation Algorithm." IEE Proc.-F1991;138;1;55-60.
SHAW IS:"Meural Network Based Controller For The Balancing Tank Of a Wastewater Treatment Plant." Journal of the SA Inst.of Chem.Eng. 1993;35;4;10-12.
SIMÕES MG, BOSE BK: "Application of Fuzzy Neural Network in the Estimation Of Distorted Waveforms" IEEE International Symposium on Industrial Electronics, June 1996; 415-420, Warsaw, Poland.
SIMÕES MG, BOSE BK: "Neural Network Based Estimation of Feedback Signals For a Vector Controlled Induction Motor Drive," IEEE Transactions on Industry Applications, May/June 1995; 31; 620-629.
STREFEZZA M, DOTE Y: "Neuro-Fuzzy Motor Controller," Proc. of IEEE Intern. Workshop on Neuro-Fuzzy Control, Mururan , Japan, 1993; 72-79.
VON ALTROCK C: *Fuzzy Logic and Neurofuzzy Applications Explained*. Prentice Hall, 1995.

WASSERMANN, PD: *Neuro-Computing: Theory and Practice.* 1989, New York.
WERBOS P: "Beyond Regression: New Tools for Prediction and Analysis In the Behavioral Sciences," Ph.D. dissertation, Harvard University, Cambridge, MA, USA, 1974.
WERTH A: "Combining Neuro-Computers and Artificial Intelligence For Real Brainpower." SA Instrumentation and Control, July 1993.
WIDROW B, HOFF M: "Adaptive Switching Circuits," Conv.Rec. WESCON 1960; Part IV;. 96-104
WILLIS MJ, DIMASSIMO C, MONTAGUE GA, THAM A, MORRIS AJ: "Artificial Neural Networks In Process Engineering." IEE Proc-D; 138; 3; 256-266.

10 PRACTICAL FUZZY CONTROLLER DEVELOPMENT

10.1 Fuzzy controller development tools

Among the fuzzy systems discussed earlier, rule-based fuzzy controllers are the most effective and practical forms of fuzzy control applicable to industrial systems. Such controllers may be implemented either in software or hardware. The question should arise as to what the trade-offs are between fuzzy implementations in software and hardware. The performance of a fuzzy controller depends, to a very great degree, on its tuning. In designing a fuzzy controller many more choices and options exist than in the case of conventional controllers. The design and optimization (i.e. *tuning*) of a fuzzy system is burdened by the many *degrees of freedom:*

$$F = k \times k_1 \times r \times r_1 \times r_2 \times m \times p \times d \tag{10.1}$$

where m = number of input variables; p = number of output variables; k = number of membership functions for each variable; k_1 = shape of membership functions for each variable; r = number of fuzzy rules; r_1 = choices of inference expressed in the fuzzy rule structure; r_2 = degree of support associated with each rule; d = choice of defuzzification method. Many of these choices are based on existing empirical data and design guidelines.

Fuzzy controller software development packages should make it practical to try out many design options in an efficient and rapid manner, either on appropriately simulated plants, or else *in situ* on the controlled plant itself. The many degrees of freedom inherent in fuzzy system design require a great deal of flexibility as regards trial-and-error and the availability of many system design options, such as, for example, easy ways to select different fuzzification and defuzzification schemes as well as aggregation and composition operators. Such flexibility suggests the use of software rather than hardware solutions. The higher speed capability of hardware-based systems is of no advantage due to the large thermal, mechanical and electrical time constants responsible for the relatively slows response of industrial plants and processes. In addition, fuzzy hardware designs cannot offer much flexibility because the chip designer must incorporate a limited choice of the design options, thereby reducing design flexibility.

Modern fuzzy controller software development systems can run on a wide choice of microprocessor platforms with cycle times of less than *100* μs which means that speed is not sacrificed for the sake of flexibility. For these reasons in recent years fuzzy *software* development systems have become the predominant design tools over specialized fuzzy controller chips. It is envisioned, however, that fuzzy *hardware chips* incorporating the most commonly used functions implemented in the form of special fuzzy commands will become popular in simple fuzzy controllers aimed at the mass market of household appliances such as, for example, washing machines, rice cookers, vacuum cleaners, thermostats, automotive components, bank teller machines, cell phones, fax machines, copiers and the like.

The charge often leveled against fuzzy controllers is that their design is *ad hoc* and no systematic design procedure exists. In fact, systematic fuzzy design protocols afforded by modern fuzzy software development packages do exist, albeit incorporating certain trial-and-error procedures as part of a cylic design procedure, described in Section 10.7 in more detail. One must not forget that even conventional design consists only of repeated analysis, that is, of a cyclic activity, as opposed to a sequential step-by-step procedure often envisioned by the less experienced. The fact is that no design procedure, however systematic, is valid unless the designer is reasonably familiar with the physical or chemical plant or process to be controlled, including its problems and vagaries. *No systematic design procedure applied mechanically to a real-life system can ever generate a satisfactory control system in the absence of any previously known information.* The more the designer knows about his system (even if only qualitatively!) the easier he will find to choose the right values for the controller parameters, and the less time-consuming tuning will be needed. It may be said that the designer's relative ignorance of the plant or process under control is traded off for a lengthier tuning cycle.

10.2 Essential features of a fuzzy software development system

In principle, a rule-based fuzzy controller can be developed and coded in any high- or low-level computer language (such as, for example, C^{++} or a microprocessor assembler language) and run in a *PC* linked to the plant under control via a serial cable. However, programming is tiresome and time-consuming, in as much as it is difficult to make program changes related to rules and membership functions. After each change the software must be recompiled or re-assembled, and even re-linked. Another problem is that it is very difficult and at times even impossible *to assess the effects of changes on the controlled plant's response.*

Thus as a minimum, a *generic fuzzy controller software development system* should make provisions for the following :

- the design of a rule-based fuzzy controller,
- the elimination or substantial reduction of the programming requirement,
- the capability of making design changes quickly,
- the availability of many options corresponding to the many degrees of freedom typical of a fuzzy control system,
- feedback to the designer to help him assess the effects of his recent modifications.

10.3 System initialization

The first step is to initialize the development system by entering

- the controller structure,
- the definitions and capabilities of input and output interfaces,
- an initial set of membership functions,
- an initial rule set

Figure 10.1 shows an example of the general control structure, that is, a number of input and output variables as well as a single rule block. The linguistic name of each input and output variable must be defined and entered. In addition, the data type of each variable (for example, *8*-bit integer, *16*-bit integer, floating point) must be selected. There should be a capability of at least *8* inputs per rule block and the use of several rule blocks. It should be possible to mix fuzzy variables with crisp ones (i.e. *fuzzy singletons*). The maximum allowable number of membership functions per variable should be at least *7*. Figure 10.1 expresses the following general rule format:

$$\textbf{IF } X = A \textbf{ AND } Y = B \textbf{ THEN } Z = C \qquad (10.2)$$

The use of more than one rule block can often greatly reduce the total number of rules, especially whenever many combinations of input variables of some rules are not expected to occur.

Example 10.1: Assume that there is a single rule block with *4* inputs and *1* output and that each input uses *7* identical membership functions. In this case, the number of unique rules would be $7^4 = 2401$ because each rule would have to express all possible combinations of memberships for each input variable. However, many of these rules would never be fired in a practical case. Assume that input variable *a* is a fuzzy singleton and thus can have only two values: *0* or *1*. Thus the total number of rules possible is $2 \times 7^3 = 2 \times 343 = 686$. Instead of a single rule block with *2* groups of almost identical rules, *2* separate rule blocks, each using one of the two values of variable *a*, would be more economical. Of course, the validity of all of the remaining system-specific combinations should be investigated in order to try to bring about further rule reductions. Designers must always consider the use of more than one rule block in order to obtain the most economical rule structure.

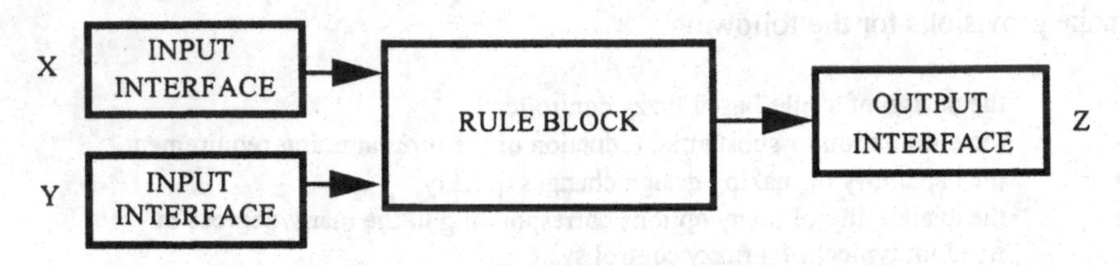

10.1. General control rule structure.

Example 10.2: Assume that the development system allows only up to *7* fuzzy membership functions per variable and that the project on hand would require as many as *19* crisp values (fuzzy singletons) corresponding to the position of a *19*-position switch each of which would use the same control strategy. The solution is to use a "*rate of position change*" instead of a "*position*" variable. The fuzzy controller output variable would thus also be a rate variable which would have to be numerically integrated with the same time constant as that of the control cycle in order to regain the position information. However, numeric integration would introduce a continuous value for the switch position which must be converted into a discrete value, for example, by means of a software-based *Schmitt trigger*, before defuzzification. The *Schmitt trigger* would also prevent any undesirable jumps in the discrete output as a result of small changes in its continuous input.

Input interface capabilities refer to the *fuzzification options*:

- no fuzzification, i.e. the inputs are already fuzzy,
- computed fuzzification,
- table-look-up fuzzification

Output interface capabilities refer to the *defuzzification* options:

- no defuzzification, i.e. output remains fuzzy
- Center-of-Area (centroid)
- Center-of-Maximum
- Mean-of-Maximum
- Bounded Sum

10.4 Graphics manipulation instead of programming

As has been stated earlier, fuzzy controllers require lengthy fine-tuning which necessitates many adjustments of rules and membership functions. For this reason, fuzzy controller software development systems should offer *graphics-based tools* which make possible *rapid changes and the observation of their immediate effects* on plant response. In this way, the designer can concentrate on the problem rather than on coding a fuzzy controller program, and his/her activities would only consist of manipulating graphic images on the computer screen. Once satisfied with the fuzzy control system's performance, he/she can instruct the computer to generate the program code automatically in one of the the languages available in the development package. If the software package runs on a *PC* using a portable C^{++} code, the entire fuzzy controller becomes a C^{++}-callable instruction with a set of parameters related to various controller properties. In turn, this instruction is incorporated in a main program which also operates the interfaces to sensors, actuators and communication channels. In addition, options to translate this code to the assembler language of the most popular microcontrollers should be available. In turn, the thus translated fuzzy controller run-time code could be downloaded to the specific microcontroller selected. See also Section 10.8. In the case of simple mass-produced fuzzy controllers, the assembler code could be downloaded in the factory to the specific product's controller board and stored in an *EPROM* used in conjunction with a microcontroller. In this way, custom fuzzy chips would be supplanted by cheaper and easily available microcontrollers. In the following, various features of a generic development system will be discussed in order to help the control engineer select the specific commercially available system that offers most of these features.

10.5 Initial membership functions

Membership functions, to be entered as a part of the input and output interfaces by means of a suitable editor program, are characterized by their linguistic names, shapes, numbers and relative positions within their universes of discourse. Options for both graphic and numeric entries as well as different colors to distinguish between them would be advantageous. A choice of standard shapes such as *trapezoidal, triangular, Gaussian,* $\cos^2 x$, *spline*, or *user-defined* should be on offer. The initial membership functions referred to would be based on the designer's initial concepts, subjects to later modification during the controller tuning or neurofuzzy training process. The choice of output membership functions (i.e. fuzzy singletons instead of functions with finite areas) would also be influenced by the defuzzification method chosen, as discussed in Chapter 8.

10.6 Rule base definition

A suitable editor needs to be available to enter rules representing various forms of fuzzy inferences such as *max-min*, *max-product*, *sum-product*, etc. for a number of input and output variables. *The inference process* consists of two phases: *aggregation* and *composition.* Aggregation refers to the logical evaluation of the inputs within one rule (i.e. the left-hand side of a rule) while composition refers to the logical combination of the outputs of all rules (i.e. the right-hand side of the rules). In using the usually standard *max-min* or *max-product* inference methods, the **IF** parts employ the fuzzy **AND** connective corresponding to *min* or *product*, while the **THEN** parts utilize the fuzzy **OR** connective corresponding to *max* or optionally *sum*. In some systems, further options are also available to represent fuzzy hedges such as, for example, *very*, *somewhat*, or *just_above*..

IF X = pos_med **AND** Y = zero **THEN** Z = pos_small (10.3)
OR
IF X = neg_med **AND** Y = pos_med **THEN** Z = zero
OR
.....

However, the standard methods do not allow to express the relative weight or relative importance of a rule within the rule set. In other words, the weight of every rule is either *1* or *0* which does not allow for fine tuning of the controller. The so-called *Fuzzy Associative Map* (*FAM*) inference feature assigns an adjustable weight between *0* and *1* to each rule, called the *Degree-of-Support* (*D-o-S*). This weight is used to multiply the aggregation result arrived at after the *min* or *product* operator was applied to the left-hand side of a rule. It is also an important component of the adaptive mechanism used in neurofuzzy systems discussed in Chapter 9. Many fuzzy controller development systems allow several input variables but only a single output variable. A rule of the following two-output structure may need special handling:

IF X = *pos_big* **AND** Y = *neg_smll*
THEN Z = *pos_med* **AND** *W* = *pos_smll* (10.4)

If the system allows only a single output, two rules with same input conditions but different outputs would be equivalent to the above structure:

IF X = *pos_big* **AND** Y = *neg_smll* **THEN** Z = *pos_med* (10.5)
OR
IF X = *pos_big* **AND** Y = *neg_smll* **THEN** W = *pos_smll*

or similarly,

IF X = *pos_big* **AND** Y= *neg_smll*
THEN Z = *pos_med* **OR** W = *pos_smll* (10.6)

and thus:

IF X = *pos_big* **AND** Y = *neg_smll* **THEN** Z = *pos_med* (10.7)
OR
IF X = *pos_big* **AND** Y = *neg_smll* **THEN** W= *pos_smll*
OR
...

However, some systems do not allow the same input conditions to generate two different outputs within the same rule block. The rule of Equation (10.4) is decomposed into Equations (10.5) and two rule blocks as shown in Figure 10.2. The rule of Equation (10.6) is decomposed into Equations (10.7) shown in Figure10.3. but it is not necessary to have two rule blocks since the truth table can have two output columns.

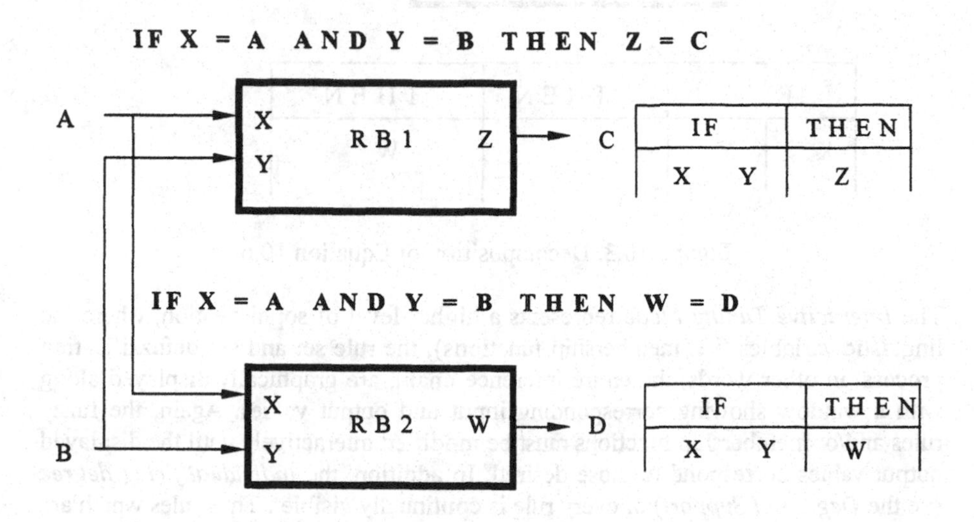

Figure 10.2. Decomposition of Equation 10.4

10.7 Tuning the fuzzy controller

The fuzzy controller software development system should offer various methods to tune a rule-based fuzzy system during the design process. This point is emphasized because once the initial rule set and membership functions have been entered, the designer has to test them in a suitable manner, with the appropriate tools provided. After the initial entry of membership functions and the rule base, various system optimization or tuning modes are required. The tuning modes described below represent some of the desirable features of a generic fuzzy controller software package.

The *File Tuning Mode* is the most obvious one: it uses a file containing sampled values of the difference between the setpoint and the process output as an input to the fuzzy controller. In turn, the fuzzy controller's output data which would feed the process, could be compared with a file containing the desired process input values. Thus the fuzzy controller would have to be modified until the output file it generates is sufficiently close to the file of desired values.

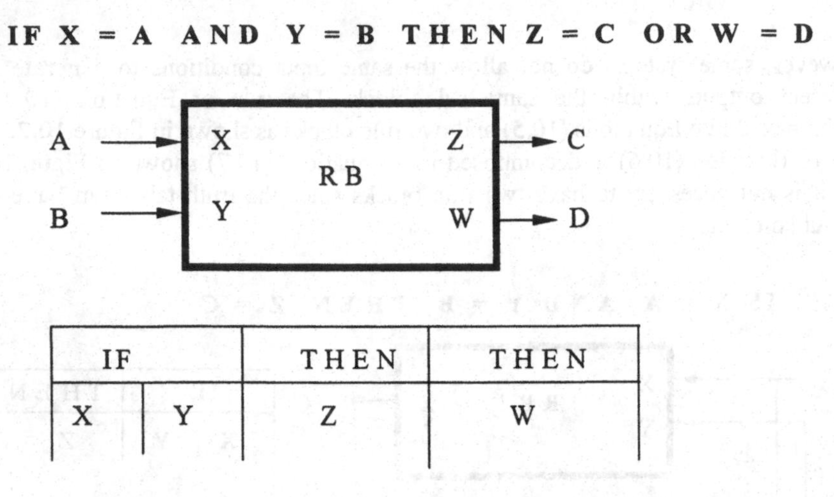

Figure 10.3. Decomposition of Equation 10.6

The *Interactive Tuning Mode* represents a higher level of sophistication, where the linguistic variables (i.e. membership functions), the rule set and the defuzzification process, in other words, the entire inference chain, are graphically displayed along with a window showing corresponding input and output values. Again, the fuzzy rules and/or membership functions must be modified interactively until the displayed output values correspond to those desired. In addition, the *individual firing degree* (i.e the *Degree of Support*) of every rule is continually visible . Thus rules which are never fired or fired only to a very small degree can be pinpointed. The development time of a controller is accelerated since the effects of design changes are immediately visualized. In this mode, *the effects of modifications to membership functions or rules are immediately discernible*. However, structural changes to the system, such as adding new interfaces, rule blocks, or new linguistic variables, are not possible. The chief advantage of the *Interactive Mode* lies in its simplicity, because it does not require the construction and programming of an approximate process model and associated graphic display like in the mode described next.

Some development systems provide an even higher level of sophistication by means of an *Off-Line Tuning Mode*. Whenever the plant or process can be simulated adequately with some kind of a model, this model can be linked up with the fuzzy software development system within the same *PC* and tuned as required. The model does not necessarily have to be a closed-form mathematical process simulation model. (In fact, if this were required then there would be no need for a fuzzy controller, for fuzzy controllers are used primarily whenever an accurate plant model is not available!). Difference equations, piecewise linear functions, look-up tables based on measurements that approximate the plant/process input-output function may be used, whereby the plant or process input-output behavior is represented accurately in at least a few operating points. One might call this model a *behavioral model* rather than a complete and accurate model. It is, however, desirable to calibrate the model at least in the few operating points mentioned.

Programming tools supplied by this type of a development system should enable the designer to *link his fuzzy controller with this plant simulation model via a software link* in order to achieve proper tuning. The plant simulation model must also include some sort of a *dynamic graphic display* such as, for example, one depicting the physical construction of the plant whose motion can be visualized. For example, the simulation model for the fuzzy controller of the container crane (see Figure 10.4) shows the movement of the crane. Alternately, the graphic display may be a replica of the instrument panel seen by the human control operator whose actions the designer wishes to incorporate into the fuzzy controller. For example, in designing a fuzzy controller for a locomotive, the graphic display would show the locomotive control panel observed by the driver. The graphic display provides *feedback to the designer about plant behavior* while he/she modifies the fuzzy rules and membership functions. Without such feedback the designer would be lost as to what modifications to effect in the fuzzy controller in order to improve its operation during the tuning process. It is rather unfortunate that some of the fuzzy controller software development systems on the market fail to include provisions for visual feedback, thereby rendering their systems impractical and useless.

Controller performance optimization is a cyclic tuning activity that consists of repeated simulation and controller modification. After such tuning, the fuzzy controller is able to control the simulated plant or process model and show its workings on the graphic display. This model would, of course, only be approximate and would not include all of the factors that might influence the real plant or process. For example, in crane simulation the effects of possible wind would not be taken intoconsideration. *Off-Line Tuning* is often followed by *On-Line Tuning* where the approximate plant simulation model that has already been tuned during *Off-Line Tuning,* is replaced by the actual plant/process and further refinements of the fuzzy controller are effected to take into consideration factors that could not fit into the initial behavioral model.

Off-Line Tuning is recommended for large-scale industrial plants and systems where good reproducibility and a rapid design cycle are of paramount importance, or where process simulation data are or can be made readily available as an aid to the contruction of a behavioral model.

Admittedly, *Off-Line Tuning* does require the writing of a subprogram for the behavioral model in a high-level language. In addition, programming the graphic display of the plant/process would constitute about two-thirds of the total effort. However, the designer could judiciously construct a sufficiently simple graphic display for the purposes of an adequate visual feedback so as to reduce the graphics programming effort. (In this respect, the graphics of the design example of a container crane shown in Figure 10.4 could probably have been restricted to only the graphics essential for showing the crane movement.)

A more sophisticated variety of *On-Line Tuning* would require a special software edition, whereby the fuzzy controller development system software running in a *PC* would be connected via a serial communication link to a remote *PC* (the target system) slated to run the actual plant/process. In using this special software, it is possible to change, i.e. fine-tune, fuzzy rules and membership functions "*on-the-fly*", i.e. while the plant is running, and thereby directly and immediately observe the effects of changes on the plant. A much less complex and less expensive but also less flexible version of this specialized software is sometimes also available to permit the designer to make some limited changes in the fuzzy controller "*on-the-fly*".

Fuzzy software development systems using *Off-Line* and/or *On-Line Tuning* can also utilize a *Neurofuzzy Option* whereby the rules and membership functions are generated automatically on the basis of on-line measurements, as described in Chapter 9.

Another version of *Off-Line Tuning* supplied by some development systems is the *Serial Link Mode*, that lets the fuzzy controller be linked to a plant simulation model located in a different hardware platform. For example, the fuzzy controller would be running in a *PC* and the plant simulation model on an external printed circuit board connected to the *PC* via a serial port and cable.

In fact, the *Serial Link Mode* can also be used in a *PC* to run in a closed-loop controlling the plant/process directly via a serial *RS232* interface. In this case, the process hardware must send input variable values over the serial link to the fuzzy controller development system and then receive the computed output variable values. Since the fuzzy controller can be modified in this mode, this functionality is similar to the "*on-the-fly*" modification feature of the *On-Line Tuning Mode*. However, in the *Serial Link Mode* the fuzzy controller outputs are calculated in the development system *PC* and sent over to the process, in contrast to the *On-Line Tuning Mode*, where the fuzzy outputs would be calculated locally on the target *PC*. Thus the *Serial Link Mode* would be strongly dependent on the communication channel quality as regards errors and response time.

10.8 Performance analysis tools

The behavior of fuzzy logic systems can be analyzed on the basis of their *input-output characteristics* and their *time response* and a fuzzy software development package should include such analysis tools.

In general, the input-output characteristics of a fuzzy system are represented by a multidimensional nonlinear hypersurface in the control space which contain all system states (i.e. operating points) determined by the input and output variables and of which only a *3*-dimensional section can be visualized. Another analysis tool is a *2*-dimensional cut of the control space constituting a *transfer plot* which displays the values of input variables on the horizontal and vertical axes, and the output values in terms of color in the region between the axes. Horizontal and vertical projections of the transfer plot might depict the variation of the input variables over a particular cross-section of the transfer plot. Any particular operating point may be selectable by means of a movable crosshair. In the case of more than two input variables, those not selected for the transfer plot would remain constant.

The purpose of the *transfer plot* is to check the *completeness of the rule base.* Operating points not covered by inference rules, or where the membership functions assigned to a variable used in a rule do not overlap, are assigned default values displayed by a uniform color like, for example, white or black. A legend translates colors to output values, including the defaults which can be read by means of the crosshair. Figure 10.8 shows a transfer plot for the dynamic simulation example of Section 10.11.

During dynamic simulation, the operating points describe a *trajectory* in the transfer plot. The trajectory shows whether or not the system states remain within the bounds of defined rules.

A unique advantage of rule-based fuzzy systems is that rules covering emergency or other unusual operating conditions can also be placed into the rule block. Such rules would be dormant (i.e. would not contribute to the controller output) until the condition arises and would only slightly increase the execution cycle time of the fuzzy rule set.

Figure 10.9 shows a *3-dimensional plot* for the same dynamic simulation example which is equivalent to the transfer plot. The main purpose of this plot is to show the relative smoothness of the control surface. There is usually a tracing feature whereby the movement of the operating point along the surface can be viewed during the tuning process. Although not shown here, a *time plot*, if available, would be useful to display the time response of selected variables during dynamic simulation.

10.9 Fuzzy controller implementation

As has been mentioned in Section 10.4, specialized editions of fuzzy controller software development systems should be available to generate the fuzzy controller in optimized assembler code for microcontroller chips manufactured by various design houses such as, for example, *Intel™*, *Motorola™*, *Microchip™*, *Siemens™* and others. In turn, the fuzzy controller software is to be downloaded to suitable circuit boards containing such microcontroller chips and other auxiliary circuits. These boards would be utilized as *embedded fuzzy controllers* in large control systems where only some functions would be implemented by fuzzy techniques. Special dedicated fuzzy software modules and fuzzy functional blocks are also available for the products of various *Programmable Logic Controller* (*PLC*) manufacturers for use in industrial automation. In fact, fuzzy control has become accepted worldwide as a valid design method and some of the largest and best-known *PLC* manufacturers such as, for example, *AEG™, Siemens™, Klockner-Mueller™, OMRON™, Allen-Bradley™, Foxboro™, Philips™, Texas Instruments™* as well as several others are marketing fuzzy control modules for their existing *PLCs*. Nowadays these products also conform to the *ISO 9000* and *IEC 1131-7* standards which are rapidly becoming internationaly approved specifications for industrial equipment and fuzzy logic design respectively..

10.10 Performance of fuzzy controllers

It is instructive to review some of the published benchmarks and associated performance indices of fuzzy controllers developed with the aid of a software development system of the kind described in foregoing sections.) Table 10.1 gives the computational speed and memory requirements for two typical industrial fuzzy control systems. Computing times are in the form of "average/worst case maximum". The memory capacity designated by *OBJ* represents the memory requirements of the fuzzy object program in the *PC* memory. As can be seen, the relatively complex parameter estimation system for anti-lock braking runs in a *33* MHz *PC* in *C*-code at a maximum cycle time of *120* μs while the same in assembler language using an *80C166* and a *80C196KD* microcontroller runs at a maximum cycle time of *140* μs and *640* μs respectively. Although not shown, using a *TMS-320* digital signal processor chip with assembler code a cycle time of *49* μs can be attained, while a relatively slow *12*-Mhz *8051* microcontroller with assembler code runs at a maximum of *4.9* ms. The popular *PIX16C5X* microprocessor with assembler code manages *2.7* ms while a *Klockner-Moeller PS4/401 Programmable Logic Controller* using fuzzy functional blocks can still run at a remarkable *1.6* ms. Any one of these exceeds by far the normal requirements of standard industrial control systems.

10.11 Simulation example: a container crane

The following example is a simplified simulation of the operation of a container crane. However, the technique is equally applicable to other types of cranes used in harbors, steel foundries and a manufacturing environment. Its purpose is to illustrate the use of a commercial fuzzy controller development software package. The control objective is to transport a container by crane from one point (the ship) to another point (the train). The difficulty of this task lies in the fact that the container is connected to the crane by cable, causing it to sway while it is being transported at high speed. Yet transport speed is of the essence because the ship must be emptied in the shortest possible time due to high mooring fees. This task requires the simultaneous on-line optimization of two nonlinear variables: *transport speed* and *swaying angle,* thus it cannot be carried out by a simple *PID* controller restricted to a single input. Although by ignoring nonlinearities it is possible to use two *PID* controllers, i.e. one for each variable, there would be no dynamic interaction between the two control loops of the kind provided by a human operator who can simultaneously and dynamically balance the two variables against one another to achieve an optimum operation.

Table 10.1. Benchmarks for fuzzy controllers.
(Reprintd with the permission of INFORM Corp., Aachen, Germany)

Processor/Controller	**Benchmarks** 5 in/out MF 20 FAM rules 2 In / 1 Out Fuzzy PI contr	 3,4,6in/5out MF 80 FAM rules 3 In / 1 Out Antilock braking
16-bit microcontroller Assembler code 80C166, 20 Mhz	Avg 0.06 ms Worst 0.09 ms 0.61KB OBJ 30 byte RAM	Avg 0.10 ms Worst 0.14 ms 0.99 KB OBJ 36 byte RAM
PC-based, C-code generation 16-bit resolution 80486SX, 33 Mhz processor	Avg 0.05 ms Worst 0.09 ms . 0.89 KB OBJ 47 byte RAM	Avg 0.09 ms Worst 0.12 ms 1.27 KB OBJ 53 byte RAM
16-bit microcontroller Assembler code 80C196KD, 20 Mhz	Avg 0.26 ms Worst 0.38 ms 0.78 KB OBJ 58 byte RAM	Avg 0.44 ms Worst 0.64 ms 1.17 KB OBJ 66 byte RAM

Other conventional solutions require highly elaborate approaches, such as model-based or state variable control. However, these need intensive engineering and hardware/software resources (and most of all, a very long development cycle) which render these techniques economically unaffordable. For these reasons, most cranes are still operated manually by human operators who can control them quite well. Yet this type of operation has a number of shortcomings, notably the safety aspects. For example, accidents are frequent because crane operators holding a remote controller in their hands and watching the crane often stumble over objects lying on the ground. Besides, the consistency of control quality depends on individual operators, wind strength and direction, maintenance problems. like rust sports on the tracks that carry the crane traveler, etc.

The simplified layout of the container crane is shown in Figure 10.4. Fuzzy control seems to be a good solution because it uses human operator experience as a basis for control system design and can cope well with plant nonlinearities. The container crane simulation discussed allows for pure manual control as well as fuzzy control that embodies the container-transport control strategy. During a simulation run, the instantaneous values of the angle, distance and motor power are displayed in real time. The fuzzy controller is working in the *Off-Line Tuning Mode* described in

Section 10.7. The graphic display shown in Figure 10.7 is part of a plant model program where the crane's behavioral model had been linked via a software link with the fuzzy controller development software during controller tuning. Thus the graphic display provides a visual feedback to the designer during the tuning process.

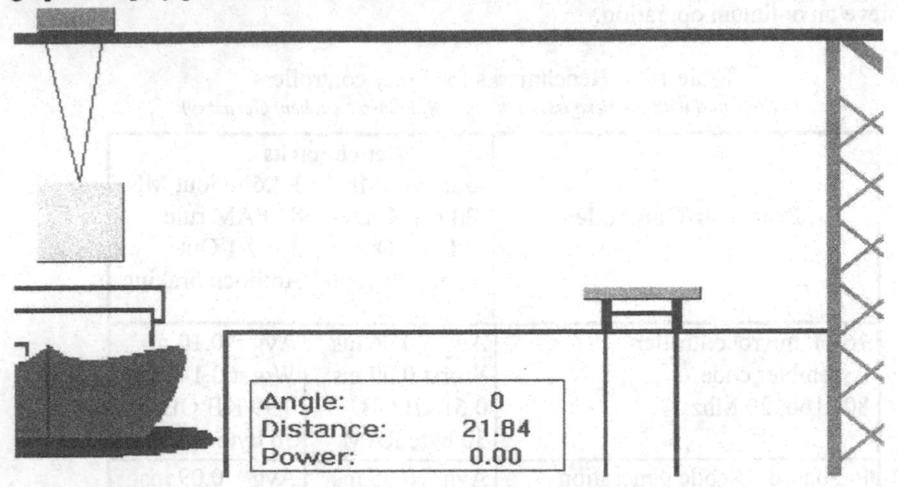

Figure 10.4. Simulation of fuzzy controller for a simplified contaner crane.
(Reprinted with the permission of INFORM Corp., Aachen, Germany)

One can readily see that a foundry crane operator would have to adopt a different control strategy if he were to transport a kettle of molten steel from, say, Point *A* to Point *B*. Instead of maximizing the speed of transfer at the expense of sway like in the case of the container crane, he would have to limit the swaying angle (to a maximum of a few degrees on each side) to avoid spillage and rather live with a slower transfer speed. Of course, the fuzzy controller designed for this control strategy would end up being different from that of the container crane[1].

10.11.1 Controller structure

The designer would enter the over-all controller structure, including the necessary inputs and outputs, into the development system by means of a *Project Editor*, as shown in Figure 10.5. In the crane example, the structure consists of two input (*Distance* and *Angle*) and one output interface (*Power*). *Distance* denotes the distance between the crane position and the target, while *Angle* signifies the swaying angle of the load. Finally, *Power* represents the power of the motor that drags the crane. The fuzzy controller itself (rule block) is the larger object connecting the interfaces. Each interface block contains a number of options which must be selected by the designer.

10.11.2 Rule structure

The rules are entered by means of the *Spreadsheet Editor*. To fine-tune the individual rules during the tuning phase of the controller design, a weighting factor, the *Degree of Support* (*D-o-S*), initially set to *1* is assigned to each rule. Figure 10.6 shows the rule set except that the bars indicating the degree of firing of each rule during dynamic simulation are not displayed whenever rules are entered into the system.

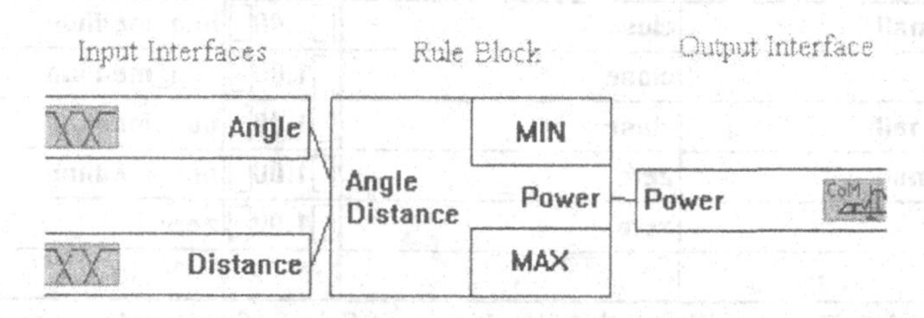

Figure 10.5. Container crane rule structure.
(Reprinted with the permission of INFORM Corp., Aachen, Germany)

10.11.3 Linguistic variables

If rules are to use linguistic terms like *far* or *medium*, these terms must first be defined. As an example, the membership functions that correspond to linguistic variable *Distance* are shown in Figure 10.7. Membership functions can be entered numerically, graphically or changed via the *Variable Editor*. More than one input membership functions as well as the defuzzification window may be displayed simultaneously by using *Tile* in the *Windows™* menu.

10.11.4 Interactive tuning mode

In the *InteractiveTuning Mode* the small arrow under the vertical bar, indicating a crisp input value for *Angle* and *Distance*, may be dragged with the mouse to another value while observing the defuzzification process (see Figure 10.7), i.e. changing the height of the output bars, while the small triangle in the output diagram (*Power*) represents the defuzzified crisp value (this cannot be dragged because it is a result). Thus it is possible to directly observe the output behavior as a result of changing an input variable.

10.11.5 Dynamic simulation of container crane.

The container crane shown in Figure 10.4 is operated by means of a fuzzy controller. As has been stated before, the control objective is to transport the load as fast as possible, even at the expense of a relatively large swaying angle. Initially to overcome the inertia of the load and get it moving, the fuzzy controller must allow a fairly large swaying angle. In turn, the angle is substantially reduced and the motion becomes fairly even until braking sets in and the load swings slightly in the direction of motion before it slowly settles down on the railroad car.

	IF		THEN	
	Angle	Distance	DoS	Power
1	zero	far	1.00	pos_medium
2	neg_small	far	1.00	pos_high
3	neg_small	medium	1.00	pos_high
4	neg_big	medium	1.00	pos_medium
5	pos_small	close	1.00	neg_medium
6	zero	close	1.00	neg_medium
7	neg_small	close	1.00	pos_medium
8	pos_small	zero	1.00	neg_medium
9	zero	zero	1.00	zero
10				

Figure 10.6. Rule set illustrating the degree of firing of each rule during a running process simulation
(Reprinted with the permission of INFORM Corp., Aachen, Germany)

If the membership functions, the defuzzification window and the crane simulation are displayed simultaneously and the crane is set in motion by activating the software link between the crane model and the fuzzy controller (in the *Off-Line Tuning Mode*), one can see the entire fuzzy inference process in a dynamic fashion as the crane moves from its initial to its final point. That is, at each instance the membership functions produce fuzzy values for each crisp input value and the rules are fired to varying extents while the fuzzy outputs are being defuzzified to produce crisp output values. Two small bars displayed around the *Degree-of-Support* in every rule (Figure 10.6), the amount of blackening indicates the extent to which the rule has been fired. The left-hand bar shows the result of aggregation whereas the right bar shows the result of composition with the *Degree-of-Support*.

10.11.6 Practical implementation of fuzzy crane controller

A *64*-ton crane that transports concrete modules for bridges and tunnels over a distance of *500* yards in a German plant was automated with the aid of the fuzzy software development system described above. The benefit gained was a capacity increase of about *20%* due to faster transportation and an increase in safety. The crane was commissioned in the Spring of *1995* and the fuzzy logic anti-sway controller has shown a high degree of acceptance by the operators.

10.11.7 Comparison of fuzzy and classical crane control

It is worthwhile to briefly summarize a detailed investigation by Benhidjeb and Gissinger [1995] of two different ways of controlling an experimental crane: optimal control and fuzzy control. The state representation of the optimal Linear Quadratic Gaussian controller was based on a state vector composed of the trolley position, load sway, cable length and their respective first derivatives. Controlling the load position with an analytic controller implies a mathematical model.

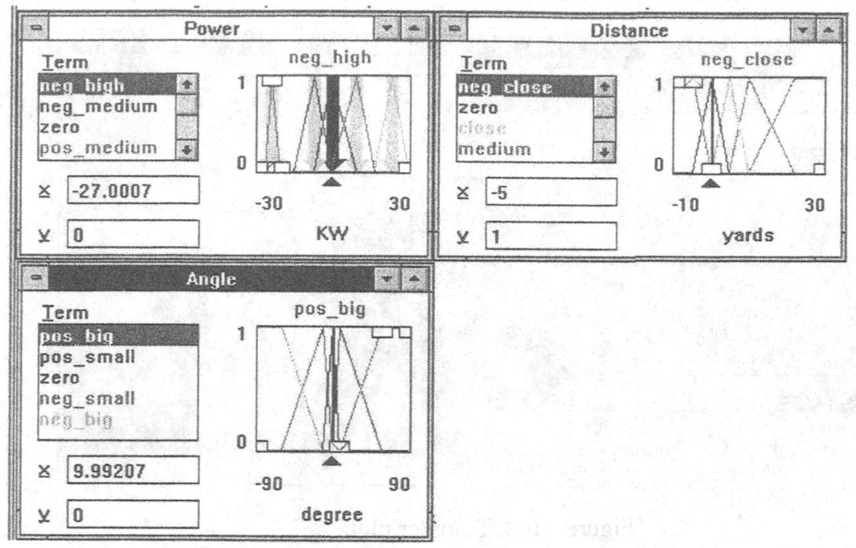

Figure 10.7. Input and output membership functions and defuzzification during a running process simulation
(Reprinted with the permisison of INFORM Corp., Aachen, Germany)

However, industrial implementation of such control without any simplification is complex and difficult, as some important variables for controlling the overhead crane are either difficult or impossible to measure. For example, angular sensors cannot be used because of their poor reliability, permanent angle offset and noise and in linearizing the system equation the cable length is assumed to be constant, which is not realistic. Besides the mathematical model for optimal load position control degrades the behavior of other state variables such as cable length and load mass. According to simulation results, a *10%* load variation degrades the results of horizontal and vertical load position, load sway, cable length and trolley position obtained.

In general, the presence of an experienced human operator was often essential to correct perturbations not predicted by the mathematical model and whenever a phenomenon could not be quantified. Examples were different forms of friction at low speeds and the condition of worn or rusty rail wheel contact points. In fact, most researchers quoted were of the opinion that the load position is not directly measurable and can only be reconstructed in mathematical models, thereby causing calculation errors. On the other hand, a human operator could give a qualitative account of a phenomenon by globally observing the environment and its tendencies to change.

A fuzzy controller embodying this human experience was expected to give better results. In fact, this turned out to be the case. The fuzzy controller was able to control the crane, even at low speeds, despite the difficulties of measuring the swaying angle and the load position. Most importantly, only the fuzzy controller could take into account the different unmeasurable and unmodelled disturbances.

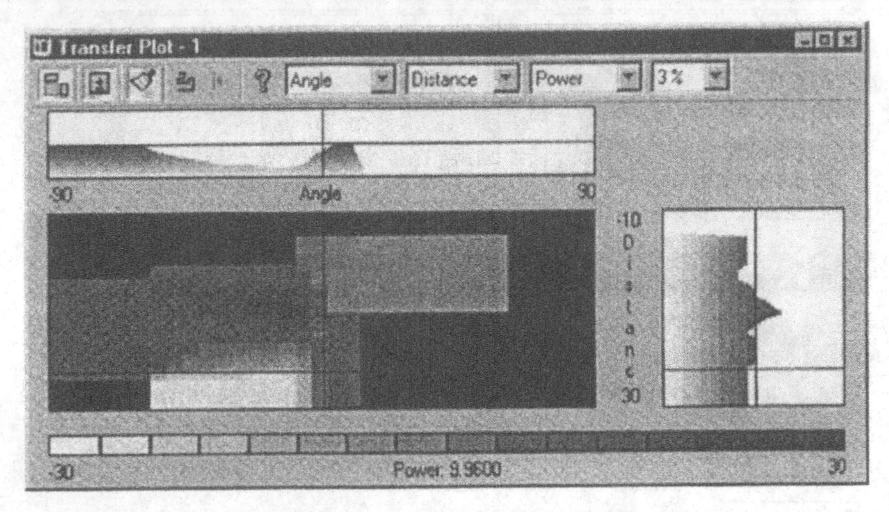

Figure 10.8. Transfer plot.
(Reprinted with the permission of INFORM Corp., Aachen, Germany)

10.12 Software design tool considerations

In a practical industrial design environment, the cost effectiveness of software design tools is of paramount importance. The high expense due to sophisticated and complex features requiring extensive training woul be justifiable only whenever a

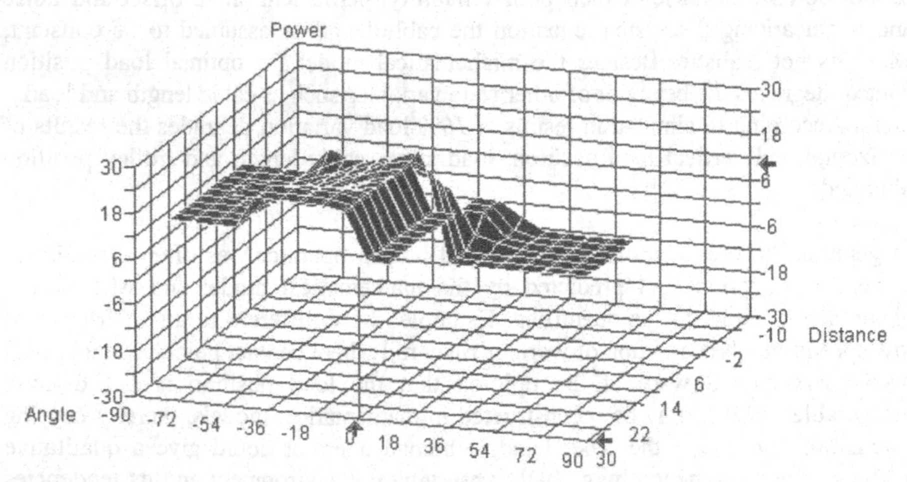

Figure 10.9. 3-dimensional transfer characteristics of crane control system
(Reprinted with the permission of INFORM Corp., Aachen, Germany)

design or modification process is needed on a recurrent basis, or whenever the design, albeit a single occurrence, would be required as a part of a large and profitable contract. However, in a tertiary educational institution the aim is to teach fuzzy design methodology and it is important to be able to demonstrate and study as many practical hands-on features of a development system as possible. In this case, the recurrent use of the software would be guaranteed by each new class of students, thereby justifying the expense, even though updating it from time to time would be highly recommended. Some manufacturers of fuzzy controller software development packages are well aware of such needs and are marketing educational packages suitable for classroom instruction. The periodic upgrading of the software for educational institutions is also being offered at a substantial discount, making fuzzy software maintenance economically feasible.

Notes

1. In running this simulation in a practical classroom session, a tried-and-proven exercise for students of fuzzy control has been to change the fuzzy controller from the container crane control strategy to the foundry crane strategy and thereby gain experience with the interactive and cyclic nature of fuzzy controller design.

References

AEG: "A Fuzzy Logic Solution Using Standard *PLCs*". SA Instrumentation and Control, Johannesburg, February, 1994.

ALLEN-BRADLEY: White Paper On Future of Automation. Milwaukee, A-B Response Center, Dept PP/0307, 107d Hampshire Ave South, Bloomington, Minn.55438, 26 August, 1995.

BANNATYNE R:"Fuzzification can be precise and deterministic".Dataweek, February 1996.

BARTOS FJ: Fuzzy Logic Reaches Adulthood. Control Engineering Online, New York, 1996; 7;1-9;

BENHIDJEB A, GISSINGER GL: "Fuzzy Control Of An Overhead Crane : Performance Comparison With Classic Control. Control Eng.Practice, 1995;3;12;1687-1696.

DRAKE P, SIBIGROTH J, VON ALTROCK, KONIGBAUER R: "Demonstration on Motorola 68HC12 MCU". FuzzyTECH Application paper, http://www.fuzzyTECH.com

FuzzyTECH-MP User Guide, Microchip Technology Inc, 1994

FuzzyTECH 5.0 *User's Manual*, Inform Software Corp. September 1997.

FRIEDEL P: "Fuzzy And Neurofuzzy Application in The Consumer Market - a Philips Perspective". Dataweek, Johannesburg, January, 1996.

FROESE T: "Applying Fuzzy Logic and Neural Networks To Modern Process Control Systems". SA Instrumentation and Control, Johannesburg, September 1994.

KOSKO B: *Neural Networks and Fuzzy Systems*. Prentice Hall, 1992

OMRON ELECTRONICS: "Fuzzy Logic: A *21*st Century Technology. An Introduction toFuzzy Logic and Its Application in Control Systems." Cat.No. FUZ-1A 11/91/3M.

PFEIFFER BM, ISERMANN R:"Criteria For Successful Application of Fuzzy Control." Engng.Appl..of Artif.Intell. 1994;7;3;245-253.

SCHWARTZ DG, KLIR GJ, LEWIS HW, EZAWA Y:"Applications of Fuzzy Sets and Approximate Reasoning". Proc. IEEE, 1994;82;4;482-497.

SHAW IS, VON ALTROCK C:"Fuzzy Logic in Industrial Automation". Elektron, Johannesburg, 15;2;12-13;1998.

SHEAR D:"The Fuzzification of DSP - Texas Instruments and Inform Software Team Up", EDN 2;9;145;1994.

SIEMENS: "Fuzzy Control: The Clear Answer to Plant Optimization Problems." Pamphlet No.E80001-V0380-A023-X-7600.

SIEMENS: "Fuzzy Control Implemented in Siemens DCS." SA Instrumentation and Control, Johannesburg, July 1993.

TAKEUCHI T, NAGAI Y::" Fuzzy Control Of A Mobile Robot For Obstacle Avoidance." Information Sciences, 1988;;45;231-248.

VON ALTROCK C: *Fuzzy Logic and Neurofuzzy Applications Explained*. Prentice Hall, 1995.
WILLIAMS T: "Alliances to Speed Acceptance of Fuzzy Logic Technology - Intel and Inform Software Team Up for 16-bit MCUs", Computer Design, 12;52-56;1992.

11 EXAMPLES OF FUZZY CONTROL

11.1 Objective

Previous chapters discussed the methodology and tools of applying fuzzy and neurofuzzy logic to industrial control problems. There are literally hundreds of successful applications of fuzzy control published in technical journals and books and the interested reader is referred to the bibliography section for their detailed descriptions. Feature articles on successful projects implementing fuzzy and neurofuzzy logic are helpful whenever the plant/process on hand is similar to those described. However, they are seldom able to furnish good general guidelines for the prospective user regarding good potential applications of fuzzy control. The objective of this section is to fill the gap in this respect.

11.2 Specific features of plants/processes to identify

A good control system designer would approach a given plant/process in an unbiased manner, seeking the most suitable control philosophy applicable to the problem on hand. In other words, he/she would have no particular preference for any specific control approach and would strive to seek the best one applicable.

There might be situations, however, when he/she would be confronted with the question whether or not an intelligent controller, such as a *fuzzy* or *neurofuzzy* controller, would be appropriate as a viable alternative. Fuzzy and neurofuzzy control might be a subject of one of the trade-off studies performed as a part of systems analysis or a proposal to another firm and in such a case its suitability would have to be analyzed in depth. For these reasons it is often necessary to critically examine a given plant/process in order to decide whether or not it is amenable to fuzzy control. What specific features of the plant/process should a designer look for in order to answer this question?

- Whenever plants/processes are complex, nonlinear, have multiple variables and are poorly understood.
- Whenever the optimization of the control system is based on technical expertise and/or the experience of a human operator rather than mathematical models; that is, whenever optimization goals include vague linguistic terms like "ease of operation", "riding comfort", or "good customer satisfaction", or the variables used are inherently fuzzy like "floor quality", "extent of soiling", "adequate softness", "material quality" which cannot be defined mathematically, or whenever a compromise between several contradictory parameters is required.
- Whenever the use of fuzzy control would provide a competitive edge such as fast prototyping, less power consumption, code size or computing speed.

An important field of application which has been on the increase and is expected to do so for sometime is *embedded fuzzy controllers* where fuzzy control would be used only for some of the functions within a conventional controller. Combining conventional and intelligent controllers in an integrated control system makes sound technical and economic sense. In this case the fuzzy controller, implemented in software, would be developed on a *PC* the way indicated previously and would be downloaded to a microcontroller board in a corresponding assembler language. In turn, this board would be built into the larger conventional controller. Table 11.1 shows the application areas where fuzzy control would achieve the greatest improvement.

11.3 Practical examples

In the following, short descriptions of some typical fuzzy control applications will be given that conform to one or more of the criteria listed above.

Video camcorder

Fuzzy controllers are applied to automatic focusing and automatic iris; the former can keep a moving subject focused continually and automatically while the latter judges whether recording is indoors or outdoors and changes the focusing velocity automatically. Human experience is necessary to estimate the quality of focusing and iris adjustment and plan the rules and membership functions at the design stage. Let us illustrate the fuzzy iris control as an example. In a strongly backlighted photo of a human figure the iris, adjusted on the basis of average light will produce a dark image of the human figure because of the undue weight given to background brightness. Dividing the frame into six parts on the basis of subject importance, adjusting the brightness based on those rankings and comparing the priorities of each zone will adjusts the iris accordingly. For example, zones *1* and *3* are given precedence if zone *2* is judged to be bright enough, and so on.

Table 11.1 Areas of improvements due to fuzzy control

SITUATIONS	IMPROVEMENTS
Compelled to operate manually	Automation
Automated, but frequent adjustments are required	Perfect automation
One input condition control	Multiple input condition control
Accurate but too sensitive	Less sensitive
Responds to incorrect manual input variations	Rssponds only to correct manual input variations
Step-by-step control of operation	Smoother operation with greater continuity
Automation combined with manual control	Automation of worker's skill to achieve total automation

Microwave oven
On the basis of information from appropriate sensors (infrared, humidity, atmospheric pressure) the fuzzy controller sets the intensity and duration of heating automatically for even very delicate cooking requirements. Human empirical knowledge about cooking and recipes is necessary to calibrate the oven at the design stage.

Washing machine
On the basis of appropriate sensors (water temperature, detergent concentration, amount of clothes, water level, fabric type, type of soiling, degree of soiling) the soaking, washing and spinning cycle time are selected automatically from among *270* possible washing cycles .Human empirical knowledge is necessary to evaluate and calibrate the fuzzy controller and to define the linguistic meaning to some inherently fuzzy variables such as type of soiling, fabric type, degree of soiling, etc. at the design stage.

DC locomotive
This is an embedded fuzzy controller developed to prevent coupler, track and motor damage during starting a heavy train as a result of uncontrolled wheel slip and incorrect taking up of train slack. It enhances and complements the existing conventional controller with intelligent functions derived from human driver/operator experience. Although recent designs use choppers and pulse-width modulated drives that incorporate rapid wheel slip control, literally hundreds of old resistor-technology locomotives needed to be upgraded in a cost-effective manner and an embedded fuzzy controller provided an inexpensive way. The fuzzy

controller is active only during the time when the *DC* motor is in a series-coupled mode, which persists until the consist is accelerated to about *25* km/h. In the parallel mode and the subsequently used weak-field mode, the fuzzy system is disconnected and the operator/driver takes over. For safety reasons, the driver can override the fuzzy controller at any time, if needed. The empirical knowledge of the most experienced human operators (i.e. train drivers) was required to calibrate and commission the system. The success of this project made it possible to extend the service life of a large number of old-technology locomotives at a low cost.

Fuzzy electrode positioning and power controller

The subject of this project was a high-power electrical arc furnace whoce control is achieved by raising or lowering the electrodes by means of *DC* motors which drive a drum-and-cable system attached to the electrodes to decrease or increase the power to the furnace. The electrode regulator consists of two main control loops: an inner position control loop and an outer slower power control loop. Considerable analytical and experimental effort was devoted to the development and tuning of the inner loop. However, the outer loop relied on a combination of discrete control (to make changes in the event the arc breaks) and *PID* control (for steady-state power regulation). Fuzzy logic was seen as the ideal control solution to provide a balance between transient and steady-state operation. The fuzzy controller is insensitive to normal process variations which occur as a result of a rapid bath surge or arc break situations. After extensive plant and controller simulation, the thus proven fuzzy controller software was installed on a *PC* interfaced with the plant via a standard bus. In turn, the controller was extensively tested and commissioned for use. In this project, the nonlinear nature of the plant was the prime motivator for the use of a fuzzy controller..

Waste incineration plant

Maintaining a stable burning temperature in waste incineration plants is important to minimize the generation of toxic gases and avoid corrosion of the burning chamber. However, the caloric value of the waste fluctuates strongly and the fire position and shape cannot be measured directly. One of the probles is the inhomogeneous quality of the refuse, while another problem is the presence of toxic emissions. A control system must keep the carbon dioxide concentration constant, maintain a uniform thermal output and ensure optimum and constant flow conditions in the furnace by optimizing combustion at a stable operating point. Conventional control methods are inadequate to cope with the inhomogenities of the feed and the concomitant variation in calorific values and ignition properties cause unavoidable variations in the combustion process and result in high emissions. Besides, much additional information, measurements and statistical data are required to determine the prevailing position and shape of the fire. The multivariable and nonlinear nature of this control problem cannot be solved by means of an exact mathematical process model. Fuzzy control reduced temperature variations to less than one-tenth of the value obtained by conventional control and also caused gas emissions to drop considerably. Human experience was necessary to validate the quality of control on site.

Air conditioner
The fuzzy controller outputs the setpoints for coolant valve, water heater valve, and the humidifier water valve. The control strategy uses different temperature and humidity sensors to determine how to operate the air conditioning process in an energy-conserving way. These comfort factors were established on the basis of human experience and judgment. This is a multivariable controller with the capability to process interdependent variables. For example, when temperature rises, relative air humidity decreases. This can be exploited by allowing the temperature controller to communicate to the humidity controller that the heater valve is about to be activated. In turn, the humidity controller can respond before actually detecting the temperature change by its sensor. The attainable energy saving and control quality of such a system is considerable, between *10* and *30* %.

Temperature control
Fuzzy logic emulates the way an expert operator would react to process disturbance if he/she were controlling the system manually. This results in much faster response to process upsets than a conventional *PID*, reduces the overshoot and allows to adjust a desired response on the basis of process knowledge. In this combination of conventional and fuzzy control, a *PID* operates during normal conditions while fuzzy control operates during an external disturbance only. The reaction speed (rise time) of the conventional controller is high but the fuzzy control cuts in at the critical time to eliminate any overshoot.

Packaging (bag sealing)
A heat seal bar was used to seal plastic bags. Such a process inherently contains an external disturbance because heat is transferred from the load (the seal bar) to the plastic bag. As a consequence, the seal bar temperature drops, but the fuzzy controller responds quickly to keep the temperature within the range needed for sealing. The uncertainties of the process and its variations with environmental factors were the prime motivation for using fuzzy control.

Carbon testing equipment:
Materials are tested for age by placing them in a centrifuge-like device and adding carbon. A constant temperature has to be maintained so that the centrifuge is surrounded by a jacket with cooling liquid circulating in it. As carbon is added, friction within the centrifuge increases, causing the temperature to rise. The fuzzy controller that regulated the cooling liquid flow quickly brings the temperature back to set point. The uncertainties of the process and its variations with environmental factors were the prime motivation to use fuzzy control.

Ore crushing mill
The objective of this project was to improve the operation of a mill used to grind pieces of rock-sized minerals down to a specific narrow size range. The mill consists of a hollow rotating drum into which the minerals are fed. Grinding occurs as a result of impact between particles caused by tumbling action as the mill rotates. Heated air with an adjustable temperature and flow rate, blown into the mill, is used to dry the minerals and to drive out the crushed particles. The saleable quality of ore

is critically dependent on the ground particle size. The grinding efficiency of the mill is governed by variables, such as feed rate, mill load, air sweep rate and temperature. The mill was originally controlled as a system of independent nonlinear processes: air temperature was controlled automatically via feedback while ore feed rate and air flow rate were controlled manually. As a consequence, process disturbances resulted in protracted periods of iterative process readjustment. The efficiency and cost-effectiveness of the mill were improved by a multi-objective fuzzy control strategy which simultaneously optimized all major control parameters.

Tension control

This fuzzy controller maintains constant tension on paper ribbon, film, steel in a rolling mill, or cable winders, despite variations in roller speed and winding bobbin diameter. A typical application is take-up machines in the manufacturing of stainless steel welding wire. As the coil winds onto the bobbin, the coil diameter enlarges, creating changes in the take-up speed and the tension. Furthermore, individual take-up machines have varying operating conditions at any given time and it is thus very difficult to control a number of them. Conventional machines use either the *pit method* which creates slack in the coil before and after the slitter, or the *dancer method* using moving rollers to adjust the tension. The *pit method* gets its name from the large pits placed before and after the slitter. The coil slack drops into these pits requiring large areas for installation. Another problem is that the coil curls in the pits and guides are necessary to align the unreeled strip appropriately for the sensors which makes it difficult to change the slit width. The *dancer roll method* employs a sensor to detect tension changes by noting position changes. A moving roller, the dancer roll, carries out tension control, usually by means of a *PID* controller. An expert operator has to set the control conditions and make adjustments on the *PID*. Although the thus attained precision is quite good, the optimum control is restricted to a narrow range which makes it impossible to respond flexibly to speed and tension variations. If the operator sets inappropriate parameters, the take-up speed varies periodically and hunting occurs in which the coil and the rolls flutter. Every time the number and/or diameters of the bobbins is changed, the system requires time and labour-consuming fine adjustments before the machine can run again (about *20* days). A fuzzy controller was installed which determines the optimum take-up speed and governs the servo motors' rotational speed to keep the tension constant. The operator inputs the difference between take-up and payout speeds, and the rate of change of that difference, as well as the dancer roll's position and its rate of change. The speed difference provides coarse control while the position difference provides vernier control. This method reduced the set-up time to *10* days and the costs of these adjustments was reduced by two-thirds. In addition, the fuzzy logic controller can control several machines simultaneously with varying tension control requirements, thus it eliminates the need for accommodating widely varying inputs and outputs and makes precise control a reality. Because it offers multiple input and output capabilities, fuzzy logic control enables the system to make comprehensive decisions and provide optimum control even for mutually contradictory events. For example, the system loads four inputs describing speed, tension and their rate of change whereupon the controller determines the optimum take-up speed for each machine even if these inputs vary in opposite directions.

Furthermore, the fuzzy controller uses rules and membership functions describing know-how, intuition and experience in software form, thus even inexperienced operators can understand the system. In fact, shop floor operators can carry out changes and readjustments with relative ease following each line changeover. The set-up procedure, used for as many as eight take-up machines, can respond to a broad spectrum of events in relation to inputs and outputs via its rules, so that it can provide precise control across a wide range of conditions with simple programming, which is easier than setting constants on a conventional system employing mathematical control techniques. Thus the fuzzy control system is able to bring the know-how of experienced human operators into sequential control systems.

Anti-lock braking system (*ABS*)
Fuzzy logic is used to optimize the existing anti-lock braking system to achieve better performance in all braking conditions. An *ABS* improves performance by reducing the amount of braking force to what the road conditions can handle. This avoids sliding and results in shorter braking distances. A fuzzy logic system estimates the quality of the road surface and can thereby improve performance. Mathematical models for braking do exist but the interaction of the braking system with the car and the road is far too complex to model adequately. Braking as a physical process can be expressed as follows:

$$s = (V_{car} - V_{wheel}) / V_{car}$$

where s = slack, $0 \leq s \leq 1$; V_{car} = velocity of car; V_{wheel} = velocity of wheel.
If $V_{car} = V_{wheel}$ then $s = 0$ and there is no braking effect. If $V_{car} = 1$, the the wheel is blocked. In between these two extremes, the braking effect as a function of slack depends on the road surface. For each of three road conditions (dry, wet, snowy) typical values of the optimum slack are *0.2, 0.12* and *0.05* respectively. Sensors that can identify road surface conditions are too expensive and not robust enough. However, the driver himself can be used as a sensor. Assume that the driver is in a car equipped with a conventional *ABS* and that at a known speed he suddenly jams the brake pedal in order to activate the anti-lock braking system. In turn, the driver can estimate the road conditions from the car's reaction. A fuzzy controller can also be installed in a car with a conventional *ABS* where the slack is initially set to *0.1* by judiciously setting the brake fluid control valves. Like the human driver, the fuzzy logic system can evaluate the car's reaction to braking and thereby estimate the road surface quality. The existing *ABS* sensors of the conventional system, such as speed or deceleration of the car and the wheels as well as the hydraulic brake fluid pressure are used. From these inputs the fuzzy controller calculates the current operating point and interpolates in its slack versus braking effect versus road surface quality table.. In turn, it adjusts the slack for the optimum braking effect. The fuzzy system tables mentioned contain the empirical knowledge gained by many years of testing on many different roads by many different human operators and in many different climates.

Cellulose production

This is an embedded fuzzy controller used during the transformation of wood to the raw material suitable to paper manufacturing. The objective is to increase the firmness of cellulose while minimizing scrap and reducing wood and water consumption. In cellulose production, wood is minced and poured into large boilers. In turn, a cooking acid (a solution of magnesium bisulphite and free sulphur dioxide) is added and the mixture is heated up to *130* C°. The reaction with the cooking acid releases lignin, leaving fibrous cellulose which is suitable for paper and textile manufacture. To produce high-quality pulp, plant operators had to balance the many different variables that influence the cooking process such as varying wood quality, cooking time, pressure, temperature, and the composition of cooking acid. This requires experience and human judgment. The fuzzy controller used incorporates the knowledge and experience of many plant operators. The controller evaluates the prevailing conditions and generates the values for optimum process control in line with human decision-making patterns. Fuzzy logic employs approximate knowledge of the plant operators to calculate the control margins for each cooking process. For example, with lower quality wood and poorer quality cooking acid, the temperature should not be as high, the mixture must be heated more slowly, and having reached a temperature of *115* C° its acid balance should not be significantly altered. Given the same lower quality wood but higher quality acid, more cooking acid can be drawn off and the mixture can be heated more quickly. Such inexact and even contradictory statements regarding input quantities that are difficult to measure and determine (such as, for example, wood quality) were previously quite unsuitable for being automated. To obtain a prescribed quality, the cooking process must be terminated at exactly the right moment. Unfortunately quality cannot be measured during the cooking process itself. It is not possible to determine whether or not the cellulose produced complies with the specifications until cooking has been completed. What was needed was a model of the cooking process which on the basis of available data could predict the necessary cooking time as accurately as possible. The old analytic model based on simplified chemical/kinetic equations, was very inexact. Neural networks proved to be the best vehicle for extracting nonlinear dependencies from stored data. With neurofuzzy control, the plant increased its monthly output by about *30%* while keeping wood consumption constant, reducing waste by three-quarters and achieving a *14%* energy savings for the same production volume.

Image processing in color copying

In terms of human perception, some of the most important factors in the color copying process are how blue the sky is, how green the leaves are, and the color of the human face. These make up what is called the "human color memory", the color brightness, chroma and hue that humans prefer to see. Since this color memory seems to be ambiguous, a better job of color balancing can be done by means of fuzzy logic. The new color balancing system takes into consideration not only the characteristics of the original to be copied, but also the average viewer's perceptions. For example, as a result of being able to define the blue appearing in fuzzy language as "a more or less bright, but somewhat lighter shade of blue", we can now get color copies of the clear blue sky that appear in the original, instead of the greenish-blue one obtained in conventional systems. Thus the fuzzy balancing system translates the brightness, chroma and hue of the original into the kind of copies that our color memories prefer.

Chemical industry
A *sewage treatment plant* and a *polymerization plant* have been equipped with fuzzy control. Since its introduction, the sewage treatment plant has used significantly less precipitation reagent resulting a rapid payback on the project. The polymerization reactor, however, provided another kind of interesting comparison between fuzzy and conventional control: the fuzzy controller operated no better the model-based conventional controller but it required only a few man-weeks for its creation, while more than one man-year was needed to build the conventional model.

Pilot projects in the chemical industry
Pilot projects in ethanol plants, butadiene plants, hydrogenation plants and ethylene oxide plants showed improvements of *5%* to *20%* reduction in variable costs. In all of these cases mentioned only a few man-weeks of work had to be invested. Although control quality of the conventional controller may have slightly exceeded that of the fuzzy controller, the latter required only a fraction of the time to develop. Experience has shown that fuzzy controllers can also replace multivariable controllers which are much more expensive to develop.

Inferential controllers
Although this section is concerned with the intelligent analysis and control of a distillation column, it is also applicable to any system wherever a conveyor belt is delivering material requiring a time-consuming real-time on-line chemical analysis of a sample of the material and a control decision must be made on the basis of such an analysis. The *inferential analyzer system* discussed below can be implemented by a neurofuzzy controller. The neural network provides the learning capability of a nonlinear function and the fuzzy controller can, in turn, carry out the control function, as required. Traditional approaches to inferential analyzers are based on linear methods such as multiple linear regression, principal component analysis and partial least squares. These methods are not appropriate for nonlinear systems. Neurofuzzy systems are capable of modeling nonlinear systems, no matter how complex, to an arbitrarily high level of accuracy. The technology is equally well-suited to batch, discrete parts manufacturing and continuous processes in a wide variety of industries. The objective of a batch distillation system is to separate the ingredients of a mixture of liquid substances by providing a temperature gradient in the distillation column. Regarding the distillation column, analysis of the top composition of the column by a chemical analyser is used to adjust the reflux, the reflux ratio or the distillate feed ratio. A problem (shared with many similar systems) is that the result of the analyser will appear after about *20* minutes after the sample is taken. The correlation between the result of the analyser and the temperature difference between the top tray and, say, the 15th tray, is unknown. However, a neural network can be trained by examples and thus information can be provided about the expected analyser results. In an actual implementation, tests have yielded

an average error of less than *0.3%* of the value, better than the theoretical accuracy of an analyser. In this way, the cost of the analyser can also be saved.

Induction motor controller

Induction motors are inexpensive, robust, reliable and highly efficient. However, they are difficult to control due to their complex mathematical model, their nonlinear behavior during saturation, and the oscillation of electrical parameters due to temperature fluctuations. For example the rotor time constant of an induction motor can change up to *70%* over a motor's temperature range. Advantages of using fuzzy logic control include short development time, easy transfer to different motor sizes, and a strong tolerance for parameter oscillations. The input variable to the fuzzy logic block was the slip frequency while the output was the stator reference current. The fuzzy control block provided a constant magnetizing current which was a nonlinear function of the slip frequency., the rotor time constant, the rotor leakage factor, and a non-constant offset current. In this version a conventional *PI* controller was used for speed control. Based on the shape of the magnetizing current and the range of motor parameters, a database was generated. In turn, the fuzzy logic system was trained by using neurofuzzy techniques to adapt the membership function and rule base to the contents of a database representing the desired system behavior. In a second, enhanced version, the *PI* speed controller was replaced by a second fuzzy controller block. The entire enhanced fuzzy logic controller was implemented in seven days. Control of an ac motor requires very fast loop times, in the order of a few milliseconds. The controller was implemented with software loaded onto a *TMS320* Digital Signal Processor (*DSP*) chip. The performance of the programmable microcontroller was as good as that of a dedicated fuzzy chip because of the pure software implementation.

Synchronous control of two conveyor belts

At an automatic packing operation in a factory, products and boxes travel along separate conveyor belts. The two belts must operate in synchronism, otherwise the products will not be boxed properly. *PID* control can achieve synchronization most of the time, but with occasional errors. This results in unboxed products and time lost due to error correction. With the addition of fuzzy control, perfect synchronization can be achieved consistenly. In this case fuzzy control is added as a supplementary controller, operating independently but in tune with the main *PID* control. The system works as follows. First, a travel speed is set for the conveyor belt transporting the products. As the operation begins, products and boxes move forward upon their respective conveyor belts. Photosensors record the timing as products and boxes flow past and the time gap is calculated. In turn, the difference between the current and the last recorded gap is determined. These two conditions (timing gap and their derivative, the timing gap change) are used to determine the proper travel speed for the box conveyor belt. Three membership functions are necessary for two inputs and one output. The rules are quite simple: the speed of the box conveyor belt is altered only when these two conditions are met: one, that timing gaps exist between products and boxes, and two, that these gaps are unlikely to decrease if left alone. In this way, the operation becomes completely automated. Thus fuzzy controllers have the ability to automate production jobs where this was not possible before by eliminating the need for fine adjustments. This would occur when a variable changes irregularly during operation, requiring human judgment to respond to the uncertain situation.

Reference

The following references represent a small selection of the hundreds of publications that exist on the industrial application of fuzzy control systems. Most of the works cited describe practical realizations rather than laboratory research results.

AOKI S, KAWACHI S, SUGENO M:"Application Of Fuzzy Control Logic For Dead-Time Processes In A Glass Melting Furnace."Fuzzy Sets and Systems, 1990;38;251-265.

BAUER V, FURUMOTO H:"Fuzzy Logic Optimizes Pulp Production." Siemens Bulletibn, Industrial and Building Systems Group, Dept ANLA74, PO Box 3240, D-91050 Erlangen, Germany.

BONISSONE PP et al:"Industrial Applications of Fuzzy Logic at General Electric".Proc.IEEE,1995;83;3;450-464.

BENHIDJEB A, GISSINGER GL:"Fuzzy Control Of An Overhead Crane: Performance Comparison With Classic Control." Control Eng.Practice, 1995;3;12;1687-1696.

ERENS F:"Process Control cf a Cement Kiln With Fuzzy Logic." EUFIT'93 Conference, Aachen,1667-1678.

GEBHARDT J, VON ALTROCK C: "Recent Successful Fuzzy Logic Applications In Industrial Automation." Fifth IEEE Internat'l Conf. on Fuzzy Systems, New Orleans, September1996..

GRAHAM BP, NEWELL RB:"Fuzzy Adaptive Control Of A First-Order Process."Fuzzy Sets and Systems, 1989;31;47-65.

HAYASHI K. et al: "Neuro Fuzzy Transmission Control for Automobile with Variable Loads"IEEE Trans.on Control Sys.Techn.,1995;3;1;49-53..

HOFBAUER P, AREND HO, PFANNSTIEL D:"New Heating System Controls Based On The Use Of Fuzzy Logic." EUFIT'93 Conf.1993;1036-1042, Aachen, Germany.

JUNG JY, IM YT, LEEKWANG H:"Fuzzy Approach To Shape Control In Cold Rolling Of Steel Strip." Electr.Lettrs, 1994;30;12;1807-88.

KOENIGBAUER R, KRAUSE, VON ALTROCK: Fuzzy Logic and TMS320 DSP Enhanced Control of an Alternating Current Motor. Application Report, Texas Instruments 1996.

KRAUSE B, VON ALTROCK C, LIMPER K, SCHAFERS W:"Development Of A Fuzzy Knoeledge-Based System For The Control Of A Refuse Incineration Plant."EUFIT'93 Conf, 1993;1144-1150.

LARSEN PM::"Industrial Applications Of Fuzzy Control."Int.J.Man-Machine Studies, 1980;12;3-10.

LEE D, LEE JS, KANG T:"Adaptive Fuzzy Control Of The Molten Steel Level In A Strip-Casting Process." Control Eng.Practie, 1996;4;11;1511-1520.

MAEDA M,MURAKAMI S:"A Self-Tuning Fuzzy Controller."Fuzzy Sets and Systems, 1992;51;29-40.

OISHI K. ert al:"Application of Fuzzy Control Theory to the Sake Brewing Process."Journ.of Fermentation and Bioengineering, 1991;72;2;115-121.

OMRON Fuzzy Temperature Controller E5AF data sheets. Cat,No. H51-E1-1A.OMRON Corporation, Temperature Control Devices Division.

PROCZYK TJ, MAMDANI EH:"A Linguistic Self-Organizing Process Controller."Automatica, 1979;15;15-30.

ROFFEL B, CHIN PA:"Fuzzy Control Of A Polymerization Reactor." Hydrocarbon Processing,JUne 1991;47-49.

SHAO S:"Fuzzy Self-Organizing Controller and Its Application for Dynamic Processes."Fuzzy Sets and Systems,1988;26;151-164.

SHAW IS, MORS W, VAN WYK JD:"Embedded Fuzzy Controller For a DC Locomotive." Trans.SA.Inst.of Elect.Eng., June 1996;65-76.

SUGENO, M.(Ed): *Industrial Applications Of Fuzzy Control*. Elsevier, 1985.

TOBI T, HANAFUSA T:"A Practical Application Ofc Fuzzy Control for An Air-Conditioning System." Int.Journal .of Approx.Reasoning, 1991; 5;331-348.

TOGAI M, WATANABE H:"Expert System On a Chip: An Engine for Real-Time Approximate Reasoning."IEEE ExpertFall 1986;55-62.

VON ALTROCK C, et al:"Optimization Of a Waste Incineration Plant Using Fuzzy Logic." EUFIT '94 Conference, 1994, Aachen.
VON ALTROCK C:"Fuzzy Logic In Automotive Engineering."Embedded Systems Conference, Santa Clara, California, 1994,1995,1996.
VON ALTROCK C, KRAUSE B, Zimmermann HJ:"Advanced Fuzzy Logic Control Technologies in Automotive Applications." Proc. of IEEE Conf.on Fuzzy Systems, 1992;831-842.
ZHANG J, RACZKOWSKY J, MAYER F:"Development of a Fuzzy Co-Processor for Real-Time Control." EUFIT'93 Conf.,1993;91-97,Aachen,Germany.
YU C, CAO Z, KANDEL A:"Application Of Fuzzy Reasoning To The Control Of An Activated Sludge Plant."Fuzzy Sets and Systems, 1990;38;1-14.

12 FUZZY MODEL OF HUMAN CONTROL OPERATOR

2.1 Human control operator activities

It has been stated in earlier chapters that experienced human control operators are able to control complex plants/processes in a satisfactory manner without having to resort to mathematical models, even if they do not possess a deep knowledge of the dynamics that they are controlling. This chapter will examine the activities of a human control operator to be modeled, discuss traditional human operator models, and present a successful fuzzy operator model based on relational equations.

In general, the human operator is confronted with a *tracking task* which may be open-loop or closed-loop. Even if he/she controls a production process based on a recipe, such as, for example, rubber extrusion and rubber milling control, or a setup and adjustment procedure such as, for example, winding tension control, jacketed tank temperature control for complex temperature patterns, videotape head position adjustment, battery fluid density adjustment, adhesive application, flow rate adjustment, and the like, he/she is still tracking a predetermined conceptually ideal time-dependent pattern.

However, the tracking activities of a human control operator are more explicit in vehicle motion control where *the human operator would be placed into a closed feedback loop.* To do this, one needs to construct a model of the human operator whose purpose would be to describe and predict the operator's behavior in controlling a dynamic process. It was found that simple, well-defined and experimentally manageable control tasks such as *tracking* are excellent vehicles for research. The operator model would describe human control operator behaviour for different kinds of *process dynamics* and *driving functions*. The model does not necessarily have to include environmental variables such as, for example, ambient lighting, temperature, altitude, vibration and operator-centered variables such as, for example, training, motivation, fatigue, etc. which must be dealt with by the judicious organization of the experiments involved.

Tracking requires a reference or target being tracked as well as a representation of the human operator's tracking action. In *compensatory* tracking, the human control operator observes on a display device the *relative* motion of the target with respect to the action marker, i.e. their *difference* (error), while in *pursuit tracking* the operator observes the absolute motion of both the target and the action marker separately. A substantial amount of research has been done with regard to systems having few controlled variables and bandwidths limited to a few Hertz. Vehicles such as automobiles, aircraft, bicycles, ships and spacecraft as well as production machines, cranes, nuclear and conventional power plants, and hazardous chemical plants fall into this category. Traditional models of a human control operator in a *compensatory tracking loop* were the *quasi-linear* model by McRuer [1965] consisting of the describing function method and crossover frequency method, while *optimal control* methods by Kleinmann et al [1971] using state variable concepts were also presented[1]. These approaches provide a valuable insight into human control behavior, albeit under certain restrictive conditions. They were developed to assess human controller capabilities and limitations for military fighter planes, helicopters, and space vehicles. An intelligent control closed-loop human operator model, based on fuzzy logic, was developed by Shaw [1990] which overcame many of the limitations of the earlier methods.

The first objective of this chapter is to develop an insight into traditional closed-loop human tracking control operator models. The second objective is to apply the general approach of fuzzy control, namely, to capture the human operator's behavior patterns in a closed-loop fuzzy operator model which can be subsequently used in an automated control system. Recall Chapter 3, where the principles of modeling (i.e. identifying) the human operator (rather than the plant/process) were highlighted.

12.2 Sinusoidal quasi-linear model

This is a modification of the describing function method used in nonlinear systems analysis. The model uses a *sinusoidal* driving function, and the input-output relationship can be considered linear for a set of fixed conditions, despite the presence of nonlinear system elements; the deviation from linearity is handled separately as shown below. The model consists of two components:

- the *describing function*, represented by a linear (fundamental frequency) output component which correlates with the sinusoidal input driving function. (In the case of a purely linear system, the describing function is identical to the conventional Laplace transfer function).
- the *remnant* represented by the output component which does not correlate with the driving function and contains noise as well as harmonics resulting from the passage of the sinusoid through the nonlinearity). The remnant is usually characterized by a power spectrum.

Refer to Figure 12.1 depicting a simple manual compensatory control system. The human operator's task is to manipulate mechanical control devices such as, for example, handles, knobs, joysticks and the like in order to minimize the error defined as the difference between actual and ideal states of the output variable.

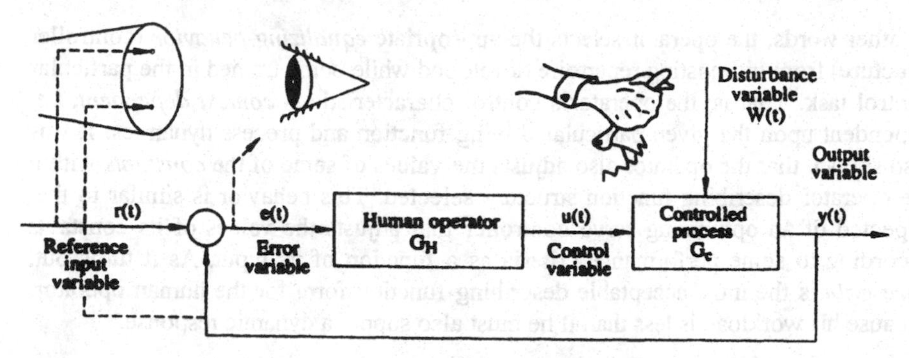

Figure 12.1 Block diagram of a compensatory tracking system

The term *quasi-linear* implies linear behavior (i.e. constant parameters for the human operator G_H for a given set of driving function *r(t)* and controlled dynamic process G_c) which ***remain unchanged within an experiment***. G_H and G_C are ***Fourier*** transforms $G_H(j\omega)$ and $G_C(j\omega)$. If the driving (or forcing) function or the dynamic process were to change, the quasi-linear human operator model would also change, but under the new circumstances it would again be linear. It was shown that by considering averaged data from many experiments, a simple describing function can be formulated whose parameters are adjustable according to certain conditions that describe the main features of human control behavior. The generalized from of the low-frequency describing function is:

$$G_H = K_p e^{-j\omega\tau} (T_L j\omega + 1) \; / \; (T_I j\omega + 1)(T_N j\omega + 1) \qquad (12.1)$$

where K_p = gain; τ = reaction time constant; T_N = neuromuscular lag, typically *0.1-0.2* seconds; $(T_L j\omega + 1) / (T_I j\omega + 1)$ = equalization characteristics which, coupled with the gain K_p, signify the human adaptability that allows the operator to control many different dynamic processes; $e^{-j\omega\tau}$ = pure time delay caused by various data processing functions of the human nervous system.

The interpretation of the *equalization term* is especially interesting. For a given driving (or forcing) function and process dynamics, the form of the equalization term is adapted to compensate for the process dynamics and the human reaction time delay. It can change to pure lag, pure lead, lead-lag or pure gain, as shown in Table 12.1 That is, operators in manual control systems exhibit the type of behavior directly analogous to that of equalizing elements inserted into a servo system to improve over-all dynamic performance! This equalizing behavior goes on continually (i.e. is time-variant), based on the data observed by the operator. It was shown experimentally that the operator adapts his/her equalizing behavior to the process dynamics and forcing functions encountered. For example, the frequency-dependent equalizing term $(T_L j\omega + 1) / (T_I j\omega + 1)$ may be adjusted intuitively by the human operator to one of the forms shown in Table 12.1.

In other words, the operator selects the appropriate *equalizing behavior* (controller structure) from his existing repertoire developed while being trained in the particular control task. That is, the operator's control characteristic is *context dependent*, i.e. dependent upon the given particular driving function and process dynamics. It was also shown, that the operator also adjusts the values of some of the *constants* within the operator describing function structure selected. This behavior is similar to that expected of an optimizing servo controller that adjusts the values of its constants according to some performance criteria as a function of its input. As it turns out, *pure gain* is the most acceptable describing function form for the human operator, because his workload is less than if he must also supply a dynamic response.

12.3 The simplified crossover model

The basic definitions of gain and phase margins are based on *Bode*'s frequency response theory. The magnitude of the gain is :

$$|G_H G_C (j\omega)| = G_H G_C / (1 + G_H G_C) \qquad (12.2)$$

and near the *crossover frequency*, ω_{cr} :

$$G_H G_C \cong 1 / (j\omega / \omega_{cr}) \qquad (12.3)$$

The *gain margin* :

$$GM = 1 / (G_H G_C (j\omega_\pi) \text{ at the phase of } arg\, G_H G_C (j\omega_\pi) = -\pi \text{ radians}$$

where ω_π = phase crossover frequency

The *phase margin* :

$$\Phi M = 180 + \arg G_H G_C (j\omega_1) \text{ degrees at the gain of } |G_H G_C (j\omega_1)| = 1$$

where ω_1 = gain crossover frequency. The simplified crossover model was developed for manual control systems according to which the human operator adjusts his describing function so that the open-loop function $G_H G_C$ in the vicinity of the crossover frequency is :

$$G_H G_C = \omega_{cr} \, e^{-j\omega\tau} / j\omega \qquad (12.4)$$

In other words, a specific process dynamic G_C "evokes" in the human operator a describing function G_H in the crossover region of the following form:

$$G_H = \omega_{cr} \, e^{-j\omega\tau} / j\omega \, G_C \qquad (12.5)$$

where the form (i.e. structure) of the human operator's response, G_H is determined by the process dynamics G_C. This crossover model is a convenient approximation suitable for engineering purposes. Although it models primarily the amplitude response, it describes the most significant features of operator behavior adequately. This is because the actual shape of the open-loop function away from the crossover frequency is usually irrelevant to the closed-loop performance. Experiments have indeed proven that the operator describing function G_H is adapted so that the open-loop amplitude response $|G_H G_C|$ tends to become $K_C / j\omega$ in the vicinity of the crossover frequency. Table 12.1 illustrates crossover models and operator describing functions for various controlled process dynamics.

Table 12.1 Operator describing functions and crossover models for various process dynamics.

Process dynamics $G_C G_H$	**Operator describing function** G_H	**Crossover model** ($\omega \cong \omega_{cr}$) G_C
$K_C / j\omega$	$K_P \, e^{-j\omega\tau}$	$K_C \;\; (K_P \, e^{-j\omega\tau} / j\omega)$
K_C	$K_P \, e^{-j\omega\tau} / (T_I j\omega + 1)$; when $1/T_I \ll \omega_{cr}$	$(K_C / j\omega) \, (K_P \, e^{-j\omega\tau} / T_I)$
$K_C / (j\omega)^2$	$K_P \, e^{-j\omega\tau} / (T_L \, j\omega + 1)$	$K_C \, K_P \, T_L \, e^{-j\omega\tau} / j\omega$ when $1/T_L \gg \omega_{cr}$

From this, the following interesting conclusions can be drawn:

- If $G_C = K_C$, i.e. a *pure gain* process (position control) then the operator assumes the describing function of $K_C / j\omega$, *i.e. an integrator*.
- If $G_C = K_C / j\omega$, i.e. an *integrator* process (speed control), then the operator assumes the describing function of K_C, *i.e. a pure gain*.

It was stated earlier that the operator has a minimum workload whenever its describing function can assume the form of a constant, K_C. The above shows that this occurs whenever the process dynamics $K_C / j\omega$,is an integrator. In the time domain, integration means rate or speed control, thus *speed control provides the minimum workload for the human control operator.*

12.4 Human Tracking Operator Model Based On Fuzzy Logic

It has been observed that human operators can successfully control complex industrial systems having dynamics that they really do not understand. With some training, the human controller can adapt his/her control strategies to the task on hand. This process of adaptation consists of learning by trial-and-error, and also referring to stored experiences. In other words, the human controller uses the observed data to create and continuously update its own "structure" which constitutes learning behavior. Handling the control task can provide solutions arrived at by using mathematical models along with simplifying assumptions to keep the mathematics tractable. However, a human control operator also performs very sophisticated tasks in an intuitive way, ranging from detection, evaluation, diagnosis, prediction and analysis to decision and selection, even though he/she may not do well in areas of calculation and quantitative processes.

Modern technology increasingly requires solutions to complex problems which cannot be formulated in a mathematical way, and which involve human experience and precedents that can only be expressed in a vague, ambiguous, qualitative i.e. fuzzy way. This is the realm of the *fuzzy logic* specifically geared to handle such information. The salient features of a machine capable of emulating these characteristics of a human operator will be *learning* (i.e. self-structuring), *estimating* (i.e. tracking) and an ability to handle fuzzy information in a systematic and rigorous manner. Such machines are called *model-free estimators*, in as much as they do not have the rigid structures found in most mathematical models. Adaptive versions build up their structure dynamically on the basis of incoming data during a *learning* or *training* phase and when the amount of learning is deemed adequate, they can proceed with the estimating task shared with conventional controllers. Fuzzy logic has the capability to translate human experience (such as, for example, control strategies) expressed in vague, qualitative, and imprecise terms into control rules. In tracking systems, the human operator expresses a whole range of qualitative, ill-defined, vague behavior patterns which are indirectly and cumulatively embedded in his input-output response to external stimuli for the given process that he is controlling. Thus the fuzzy approach concentrates on modeling fuzzily the operator's aggregate response, *investing the model with human-like intelligence.*

12.5 Fuzzy operator model characteristics

The theory of fuzzy sets lends itself well to the construction of a human operator model because it *emulates many facets of the human thinking process.* Like the crossover model, the fuzzy operator model is also *context-dependent,* i.e.

performance depends upon the *driving function* and the *plant/process dynamics* being controlled. In other words, as in the case of human operators, the form or structure of the human operator transfer function, adapted by the operator, is virtually "evoked" from the available repertoire by the specific controller process dynamics and the driving function used. Furthermore, in contrast to the other models mentioned , the fuzzy model may be equipped with *learning* capabilities, whereby the model is being built up and updated during the learning phase on the basis of the data alone. Fuzzy models are *nonlinear*, thus they have the ability to successfully emulate nonlinear dynamic systems. In addition, they have *small relative error,* have *good noise rejection* capabilities and can be realized by *either software or hardware* .

12.6 Fuzzy model identification and validation

The relational equation based fuzzy modeling technique, used to identify the discrete-time fuzzy model of the human control operator, was explained in detail in Chapter 6. A shortened explanation will be given here.

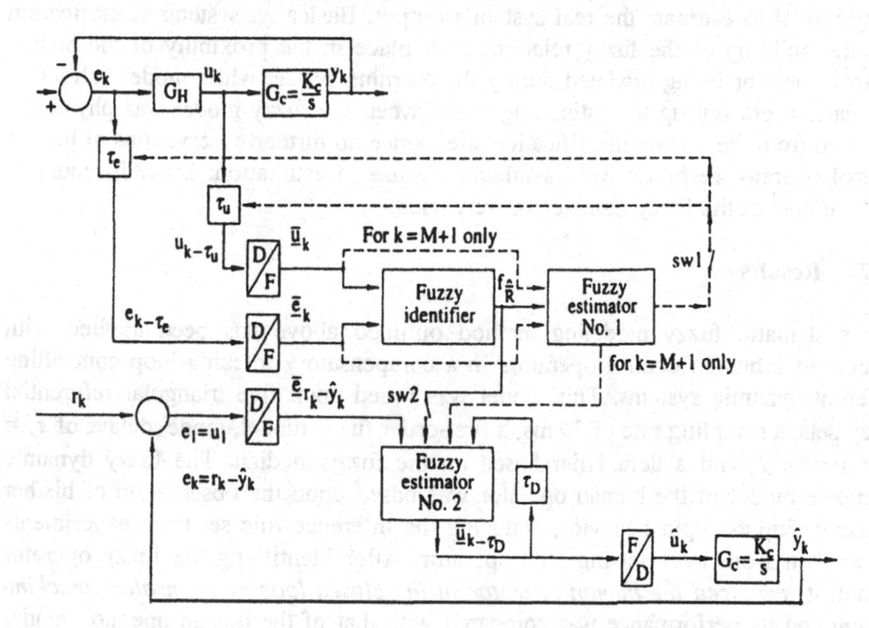

Figure 12.2. Fuzzy learning model of a human operator in a closed loop.

Figure 12.2 depicts the fuzzy learning model of a human operator in a closed loop. Assume that during the learning phase, the fuzzy relation $\mathfrak{R}$ has already been charged up with the fuzzified discrete input and output vectors $\boldsymbol{u}_k$ and $\boldsymbol{y}_k$ respectively. The process dynamics was taken as a single integrator. Fuzzy estimator No.1 which also incorporates its own fuzzifiers, is used to develop a performance index *J*, used to heuristically optimize τ_u and τ_e . During this time, software switch *SW1* is closed and *SW2* is open. Heuristic optimization consists of assuming a certain combination of τ-values in ***e[k - τ_e]*** and ***u[k-τ_u]*** and calculating the performance index *J*. This must be continued until a minimum *J* is obtained. (Naturally, if conditions do not change, there is no need to repeat this heuristic optimization procedure in every run.) In turn, software switch *SW1* is opened and *SW2* is closed and the fuzzy relation $\Re$ is calculated and delivered by the fuzzy identifier to fuzzy estimator No.2, which functions as a *recursive*[2] estimator. In turn, the estimated fuzzy state vector is defuzzified and the output of the defuzzifier delivers the estimated discrete vector to the discrete non-fuzzy process dynamics.

The fuzzy operator model was validated by applying the same stimulus to the human control operator and to the fuzzy model. In response to this stimulus, the model was required to produce the same or nearly the same response as the human control operator. Furthermore, in applications such as, for example, robot vehicle systems trained by a human operator, the validated fuzzy model was *removable from the site of the system* it aimed to emulate, due to the recursive generation of the fuzzy output. The model was put to work away from the real system (in our case, the human operator training it), and the thus learned system behavior had equipped the fuzzy model to estimate the real system's output. Basically, systems identification (i.e. the building of the fuzzy relation) took place in the proximity of the human control operator being modeled during the learning phase, while model validation was carried out during the estimating phase when the fuzzy model was physically removed from the system identification site, hence no further observations of human control operator response were available for use in estimation. Instead, recursive applications[2] of the fuzzy state vector were used.

12.7 Results

The systematic fuzzy modeling method outlined above has been applied with success to a human control operators in a compensatory tracking loop controlling different dynamic systems. This model was based upon five triangular referential fuzzy sets, a sampling rate of *30* ms, a first-order fuzzy model, model delays of τ_u = *2* and τ_y = *1*, and a defuzzifier based on the fuzzy median. The fuzzy dynamic inference model of the human operator was based upon the observation of his/her measured input-output behavior, deducing the inference rule set from experiments rather than from interviewing the operator. After identifying the fuzzy operator model, it *replaced the human operator in the closed-loop compensatory tracking system* and its performance was compared with that of the human operator under various conditions of driving functions and process dynamics.

A number of different *random-appearing* and *non-random appearing* driving functions, typical of those encountered by a human operator in a control task, were used. Random-appearing means that the driving function is unpredictable by the human operator. If the driving function were not random-appearing (such as, for example, in tracking a simple periodic sinusoidal function), then instead of following the driving function the operator would tend to anticipate it because of its repetitive nature and would synchronize his/her response with it. This would be a higher-order behavior which would complicate the simple model used. However, tracking non-random-appearing driving functions are still important in many cases. For example, Poulton [1974] discusses many experiments using ramp tracking with velocity, acceleration and jerk ramps. In most of these experiments the ramp tracks had a constant velocity. As an example, ships and aircraft usually move at a constant rate and produce constant rate tracks on a map display, although not when they are viewed directly by a fixed observer. From the viewpoint of a fixed observer, ships and aircraft moving at constant rates accelerate as they approach and decelerate as they depart. The theoretical importance of ramp tracks with constant rates is that they present to the human operator a constant unchanging problem. Changes in his/her response must therefore be determined by his/her own limitations and by the strategies which he/she uses to overcome such limitations. Experiments with ramp tracks can thus indicate the nature of the operator's limitations and strategies. Additional comments about rate anticipation and its compensation and a detailed description of the experimental results are given by Poulton.

A number of different *process dynamics* were used, such as single-integrator, constant, double integrator, first-order, and square-root. According to McRuer, using a random-appearing driving function and a single-integrator process, the human operator/controller behaves primarily like a *constant*. On the other hand, using a random-appearing driving function and a constant process, the human operator/controller behaves primarily like a *single integrator*. The fuzzy model's behavior under these conditions was compared and successfully validated with the human operator's behavior under the same conditions.

12.8 Intersubject variability

Learning implies that the fuzzy operator model was built up in real time *on the basis of measurements* on a human subject. This brings up the question of *intersubject variability* in conjunction with the quasi-linear model. In other words, let a *precise* human control operator and a *sloppy* one be given to drive, for example, an automobile. In general, a good driver strikes a compromise between precision and minimum effectiveness in order to conserve his energies and prevent fatigue. In other words, a sloppy operator in this control task is not necessarily a bad operator; in fact, as long as he/she keeps the automobile on the road without causing an accident or violating any traffic rules, he/she will not get tired as easily as a precise operator handling the same task. McRuer selected a group of highly trained and experienced operators as test subjects, and constructed the quasi-linear operator model and crossover model on this basis. Evidently, these models are going to do a better job in emulating the precise rather than the sloppy operator.

Let us look at the closed-loop responses of such a precise operator driven by a McRuer-type function (ω_c = *1.5* rad/s), and using a single-integrator process. The driving function versus precise operator output u is shown in Figure 12.3. A precise operator attempts to track the driving function to even the smallest detail, that is, his/her frequency response is better than that of the sloppy operator whose corresponding results are shown in Figure 12.4. Let us now consider the fuzzy model's emulation of the two operators' outputs. Figure 12.5 illustrates the precise operator's output u and its fuzzy emulation $\hat{u}$, while Figure 12.6 depicts the sloppy operator's output u and its fuzzy emulation $\hat{u}$. In both cases, the fuzzy model's emulation is very good indeed: the fuzzy model is capable of emulating the smaller, more precise details of the precise operator's response, but it will also track well the sloppy operator's less detailed response. Figures 12.5 and 12.6 clearly indicate the capability of the fuzzy modeling technique to learn successfully from the behavior of the different types of human operators performing the same task. This is in contrast to conventional modeling methods which require a highly selected group of human operators to validate the model. The properties of precision and sloppiness are clearly visible from the figures shown. The sloppy operator's response exhibits considerable non-linearity, i.e. both amplitude and frequency limiting.

12.9 Summary

This chapter presented the application of fuzzy logic to a human control operator model in a compensatory tracking loop. The model was shown to be context-dependent, to have learning ability, to be nonlinear and to be able to function in real time. The reference works also show that it exhibits a small relative error and model bias. Context-dependence, similar to that of the quasi-linear model of McRuer, relates to the model's ability to emulate many facets of the human thinking process.

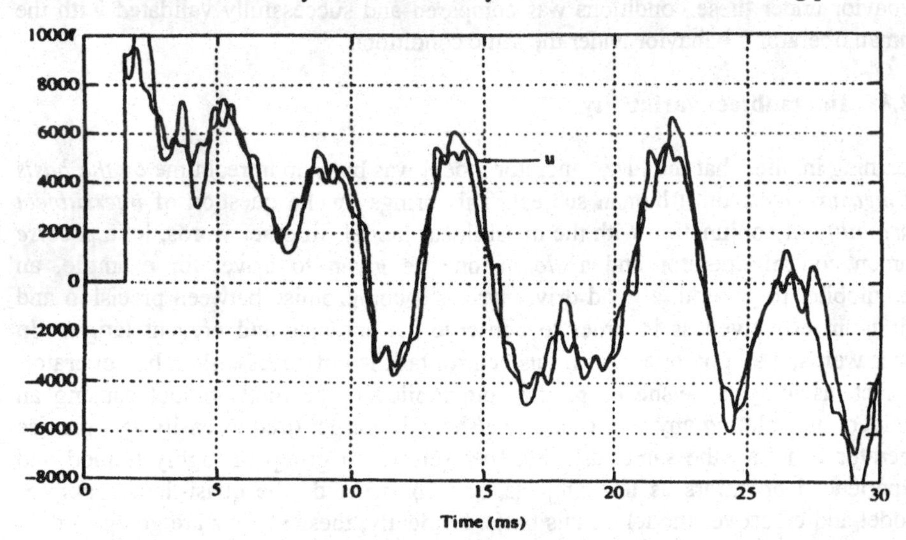

Figure 12.3. Precise tracking operator response u and its quasi-linear emulation.

As regards intersubject variability, the conclusion was that a sloppy operator who nevertheless meets the minimum performance requirements and has the added advantage of less fatigue than the precise operator, could be emulated quite well by the fuzzy model, whereas the quasi linear model has no learning ability and it's emulation of both the precise and sloppy operators is rather poor.

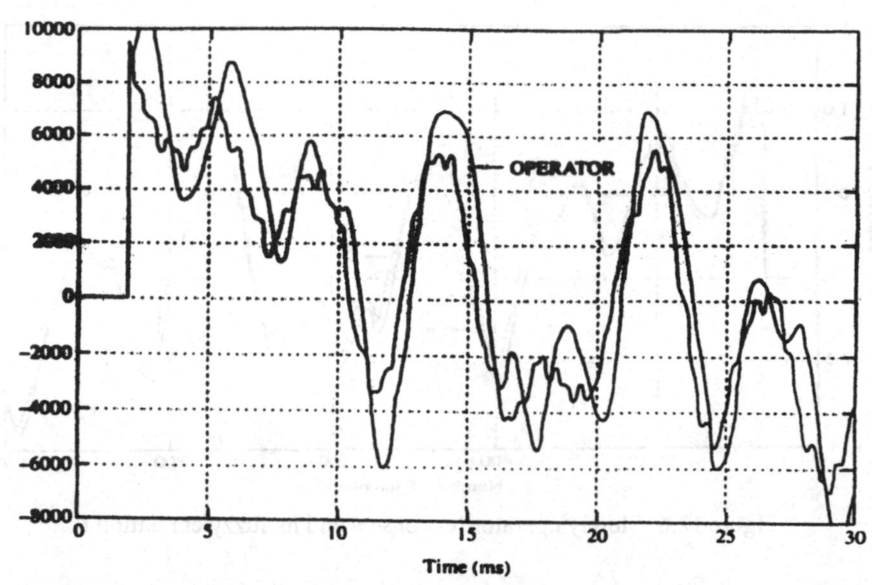

Figure 12.4. Sloppy tracking operator response and its quasi-linear emulation

The fuzzy model is able to handle various random-appearing composite sinusoidal and non-random appearing composite ramp driving functions as well as various different linear and nonlinear process dynamics. Finally, in contradistinction with the quasi-linear model which required the best trained operators as test subjects, the fuzzy model can successfully emulate any human operator, whether or not he/she is a highly experienced and trained one, or one with only average controlling abilities.

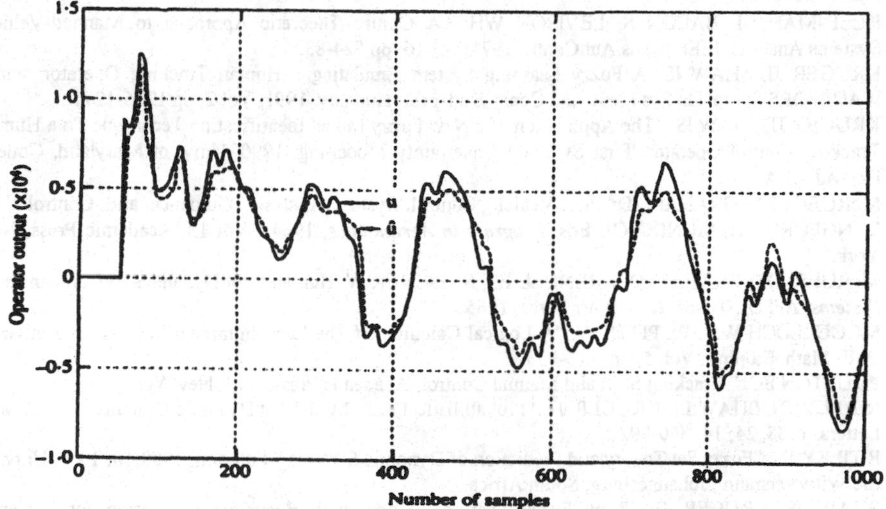

Figure 12.5. Precise operator response u and its fuzzy emulation $\hat{u}$.

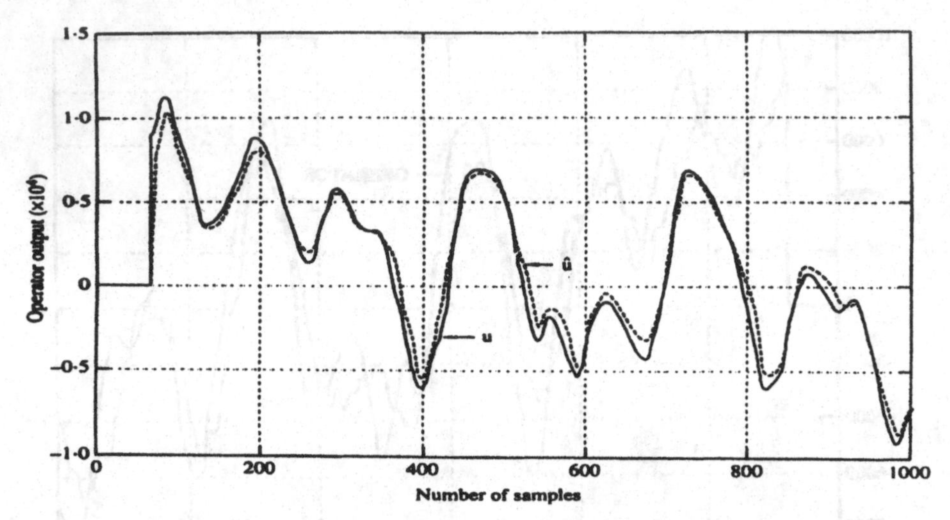

Figure 12.6. Sloppy operator response u and its fuzzy emulation $\hat{u}$.

Notes

1. The author is aware of other important contribution to the field, notably the optimal control models of Kleinmann and Wierenga. The reason for concentrating on the work by McRuer was the fact that there seems to be a certain conceptual similarity between the results of McRuer and the fuzzy operator model.
2. The recursive estimator was discussed in Chapter 2. For further information not discussed there, see the references.

References

KLEINMAN DI, BARON S, LEVISON WH: "A Control-Theoretic Approach to Manned-Vehicle Systems Analysis. IEEE Trans.Aut.Contr., 1971;Vol 16, pp 824-832.

KRUGER JJ, SHAW IS: A Fuzzy Learning System Emulating a Human Tracking Operator. Ninth IFAC/IFORS Symp. On Sys.Ident. and Contr. Budapest, Hungary, 1991, Vol 2, pp 1266-1271.

KRUGER JJ, SHAW IS: "The Application of a New Fuzzy Model Identification Technique To a Human Tracking Control Operator."First Symp.on Uncertainty Model-ing, 1990, Univ. of Maryland, College Pk, Md, USA.

McRUER DT, GRAHAM D:"Pilot-Vehicle Control System Analysis, Guidance and Control." In: LANGFORD RC, MUNDO CJ, Eds. *Progress in Aeronautics*, 1964, Vol 13, Academic Press, New York.

McRUER DT, GRAHAM D, KRENDEL ES, REISENER W: Human Pilot Dynamics in Compensatory *Systems*. AFFDL-TR-65-15. US Air Force, 1965.

MCCULLOCH WS , W. PITTS W: "A Logical Calculus Of The Ideas Imminent In Nervous Activity," Bull. Math. Biophys. Vol. 5, pp. 115-133.

POULTON EC: " Tracking Skill and Manual Control. Academic Press,1974, New York.

RIDLEY JN, SHAW IS, KRUGER JJ: "Probabilistic Fuzzy Model for Dynamic Systems." Electronics Letters, 1988; 24; 14; 890-892.

RIDLEY JN: "Fuzzy Set Theory and Prediction of Dynamic Systems." Fulcrum, 1988;18; 1 - 5. Univ. of the Witwatersrand , Johannesburg, South Africa.

SHAW IS, KRUGER JJ: "New Fuzzy Learning Model with Recursive Estimation for Dynamic Systems." Fuzzy Sets and Systems, 1992; 48; 217-229

SHAW IS, KRUGER JJ: "New Approach to Fuzzy Learning in Dynamic Systems." Electron.Lettrs, 1989; 12; 25; 796-797.

SHAW IS: "Fuzzy Model of a Human Control Operator in a Compensatory Tracking Loop." Int. Journal of Man-Mach. Studies, 1993; 39,; 305-332.

SHAW IS, KRUGER JJ: *A Computer Based Sampled Data System For The Investigation Of Human Control Operator Behavior In A Compensatory Tracking Loop*. Monograph, 1989, Rand Afrikaans University, Cybernetics Laboratory, Johannesburg, Republic of South Africa.

SHAW IS: *Expert Fuzzy Control Based Upon Man-In-The-Loop Identification*.Doctoral Dissertation,1990, Rand Afrikaans University, Johannesburg, Republic of South Africa.

SHAW I:"VLSI Hardware Realization Of Self-Learning Recursive Fuzzy Algorithm".Trans.os SA Inst.of Electr.Eng.,JUne 1995;73-86.

SHERIDAN TB, FERREL WR: "Man-Machine Systems Information, Control and Decision Models of Human Performance." The MIT Press, 1974.Cambridge, Mass.USA.

SUTTON R: *Modeling Human Operators in Control System Design.*. Res.StudiesPress Ltd,.,Somerset,England, John Wiley & Sons, 1990.

WIERENGA RD:"An Evaluation of a Pilot Model Based On Kalman Filtering and Optimal Control." IEEE Trans. on Man-Mach.Syst. 1969;MMS-10;4.

13 COLLABORATIVE INTELLIGENT CONTROL SYSTEMS

13.1 Introduction

The purpose of this chapter is to introduce the reader to some novel applications of intelligent systems. In this context, fuzzy, neural and neurofuzzy systems that possess the ability to execute learning and estimating tasks belong to this category. We have seen in Chapter 9, that in a neurofuzzy system the membership functions and rules of an initial fuzzy system are modified on the basis of measurements, or a table of measured values, for each input and output variable. However, this adaptation of the fuzzy controller takes place only once and if conditions change, the degree of fit will necessarily deteriorate. Even in the case of the relational equation based self-learning fuzzy controller where after training the system was physically removed and new estimated outputs were based on recursive techniques, the system must eventually return to the place of training to be re-trained because of the performance deterioration due to the lack of new data. An adaptive system with *periodic learning* (updating) would be desirable whenever significant changes in operating conditions are to be expected.

13.2 Concepts of collaboration

In periodic updating, whenever the environment informs the controller (whether by periodic sampling or some kind of a computer interrupt-type system) that conditions have changed and it is time for another updating, it may be said that the environment "collaborates" with the system to keep the operational parameters in equilibrium and thereby keep the operation at an optimum level. However, if only one controller system were used, no estimating (i.e. no control action) could take place while the system is learning, because there would be no controller output to the plant.

Another kind of collaboration exists between two identical intelligent systems, where one of them learns the changed, i.e. most recent, conditions while the other one performs estimating (i.e. it controls the plant on the basis of a former learning) and at a certain instant they swap tasks: the system that learned the most recent conditions carries on with estimating (i.e.the control task) while the former controlling system reverts to the mode where the most recent conditions can be learned.

A third kind of collaboration exists between a system which had learned from a human operator, and the human operator that had "taught" it. Originally the human operator was performing the estimating (i.e. control) functions while the system was learning them in the background. After awhile, the system thus "taught" can take over the control functions. The reasons for this might be, for example, that the human operator is busy with several other control functions (see Human Control Task B in Figure 13.2) and from time to time becomes overloaded. If conditions change, the human operator can again take over for awhile, while the system would upgrade its "knowledge" in the background. Clearly the two systems are collaborating to keep the control operation optimal and in equilibrium.

13.3 General criteria

Figure 13.1 shows the block diagram of a collaborative system. As stated before, each component has a *learning* and an *estimating* mode. However, a criterion for the swapping of modes is to be established. In addition, to be meaningful, a minimum amount of learning needs to be established to make sure that switching to an insufficiently trained system is inhibited. In addition, a mechanism is needed whereby the amount of change in the conditions can be determined (i.e. a certain adjustable threshold is set) and swapping is inhibited until this threshold is exceeded in order to prevent excessive swapping.

13.4 Tandem collaborative system

There have been cases for one-time training, similar to a neurofuzzy system, but using only a neural network. A typical example is the development of a neural network based hybrid controller for the compensation of distortion in electric power networks. The amount of compensation was brought about by switching in certain filters and compensators on the basis of conditions sensed on the power line. The compensation was also taking the losses into consideration which were due to the switching-in of compensator elements. The strategy was aimed at achieving minimum loss due to the switching-in of reactive elements as compensators, thereby creating an optimally cost-effective compensation.

In principle, this can be extended to the case where changes in reactive current, harmonic distortion and load conditions are used to *periodically update the control strategy* by retraining one neural network with the changed line disturbance parameters, while another *network would use the previously established compensator parameters.* Swapping of the two networks would be determined by experimentally determining the reactive current, harmonic distortion, load and their

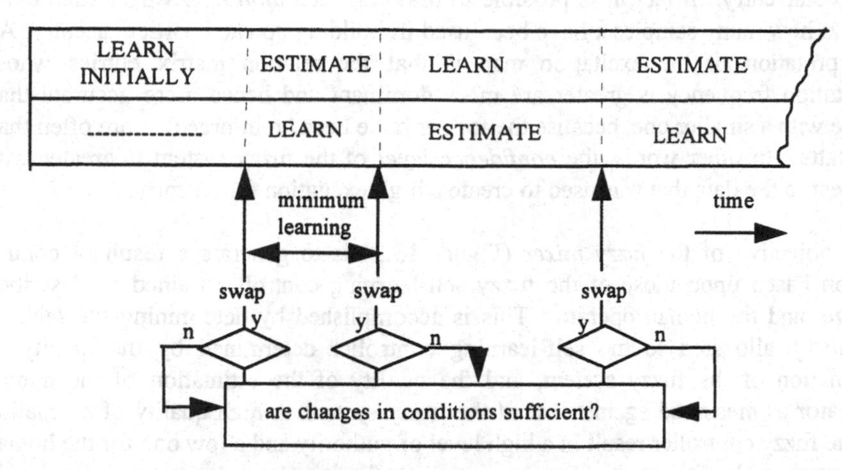

Figure 13.1. Block diagram of collaborative system.

percentual changes necessary to activate a trigger signal. The minimum necessary training time for the neural networks must be established for default conditions. Although not currently supported by a sofware design package, periodic retraining of a neurofuzzy controller for this purpose is theoretically feasible.

13.5 Collaboration between a self-learning fuzzy system and a human control operator

The implementation examples given here represent collaboration between a human operator-trained self-learning fuzzy controller and the human operator who periodically retrains it (in the background) whenever necessary.

13.5.1 Collaborative control of nonlinear multicapacity plant

An implementation by Kruger and Alberts [1993] of the collaborative control of a nonlinear two-tank plant is described as follows. The self-learning relational equation based fuzzy system, used for establishing a fuzzy model for a human control operator (see Chapter 12) was employed to be trained by manual control actions of the human operator (Figure 13.2, Human Control Task A) for controlling the liquid level in an interacting nonlinear multicapacity plant consisting of two tanks. A *fuzzy mixer* was used to combine the human operator's control actions u^h and the trained fuzzy controller's control actions u^m . Refer to Figure 13.2.

First, the self-learning fuzzy system had to be trained during its learning phase with the human operator's control actions (Task A) by calculating each element of the fuzzy relation matrix and updating it with every new incoming data item. The concept of *excitation frequency* of any particular entry of the fuzzy relation matrix was defined as the cumulative sum of contributions of each discrete sample k to that particular entry. In fact, it is possible to draw an *excitation map* where each entry shows how many samples k have been used in building up each particular entry. An interpretation of the excitation map is that the relation matrix entries whose excitation frequency is greater are more dominant and hence more accurate than those with a smaller one, because the former have been "reinforced" more often than the latter. In other words, the *confidence level* of the fuzzy system is greater with respect to the data that was used to create a high excitation for an entry.

The objective of the *fuzzy mixer* (Figure 13.2) is to generate a resultant control action based upon those of the fuzzy self-learning controller trained as described above, and the human operator. This is accomplished by determining the *level of authority* allocated to the self-learning controller determined by the quality of estimation of the fuzzy system, and the quality of the estimation of the human operator as measured against that of the fuzzy system. A high quality of estimation of the fuzzy controller result in a high level of authority and a low one for the human operator, and vice versa. Fuzzy system authority depends on the previously described confidence levels defined by the excitation frequency but can also be adjusted to some extent by adjusting the size of the learning window of the moving average filter in the front end of the fuzzy controller, i.e the number of samples entering the fuzzy system, which regulates self-learning. The collaborative controller may thus be used to strongly aid the human operator under known circumstances, and to learn without interfering whenever the human operator is performing new control strategies not yet learned by the fuzzy system. In the nonlinear two-tank system used as an experimental example, the collaborative controller resulted in significantly better over-all performance and the human operator was only active for *6%* of the time implying a reduced operator load, compared to *33%* when not using collaborative techniques. Various human operators with different control strategies were successfully emulated by the fuzzy controller for validation purposes.

13.5.2 Collaborative control of automatic automobile transmission

This example is another implementation of collaboration between a human operator and a previously trained fuzzy controller. The control strategy alternatives of automobile transmissions are contradictory. On the one hand, maximum fuel efficiency demands that one select the next higher gear as early as possible. On the other hand, maximum performance requires that one switch to the next higher gear as late as possible. With a standard shift, the best strategy is chosen by the driver on the basis of prevailing traffic conditions. However, an automatic gear box has no input from traffic conditions and the driver's wishes. The fuzzy automatic transmission controller avoids unnecessary shifting on winding on hilly roads. It also senses whether the driver wants economical or sporty performance. Finally, it avoids unnecessary downshifting to the next lower gear if the thus gained acceleration is too low. The fuzzy controller evaluates the current speed and it also analyzes driver's particular style of acceleration and braking. As regards braking, for example, it records at the number of accelerator pedal changes within a certain period. The variance of these changes is an input to the fuzzy controller which draws the following conclusions with respect to driving conditions:

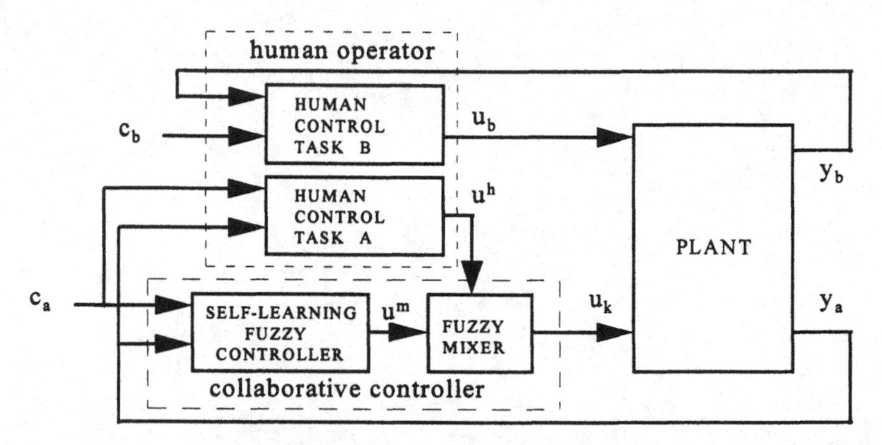

Figure 13.2. Human-machine control system with collaborative control loop.

- Many pedal changes within a period indicate a fast winding road while few pedal changes and a low variance of pedal changes indicate a freeway.
- Many pedal changes and a very high variance of pedal changes indicate a slow winding road while medium variance of pedal changes indicates a fast winding road.

This application uses *the driver as the sensor* for the driving condition. The fuzzy logic controller interprets the driver's reaction to the driving condition detected and adapts the car's performance accordingly. It tries to analyze the driver's relative satisfaction with the performance. If not, it adapts itself to suit the particular driver's needs. In other words, the actual performance is "customized" to the desires of the particular driver.

References

IKEDA H et al.:"An Intelligent Automatic Transmission Control Using a One-Chip Fuzzy Inference Engine." Internat'l Fuzzy Sys.and Intelligent Control. Conf.,1992;44-50.

PRETORIUS RW, SHAW IS, VAN WYK JD:"The Development of a Neural Network Based Controller for the Cost-Effective Operation of a Hybrid Compensator For the Compensation Of Distortion In Electric Power Networks. Proc. AMSE Conf on Intell.Syst. 1994;223-235, Pretoria, South Africa.
KRUGER JJ, ALBERTS HA:"Fuzzy Human-Machine Collaborative Control Of A Nonlinear Plant." IFAC World Congress, 1992:8;337-342, Sydney, Australia..
SHAW IS, KRUGER JJ: "New Fuzzy Learning Model with Recursive Estimation for Dynamic Systems." Fuzzy Sets and Systems, 1992, Vol 48, pp 217-229
SHAW IS: "Fuzzy Model of a Human Control Operator in a Compensatory Tracking Loop." Int. J. Man-Mach. Studies, 1993, Vol 39, pp 305-332.
SHAW IS, KRUGER JJ: “"A Computer Based Sampled Data System For The Investigation Of Human Control Operator Behavior In A Compensatory Tracking Loop. Monograph, 1989, Rand Afrikaans University, Cybernetics Laboratory, Johannesburg, Republic of South Africa.
SHAW IS: "Expert Fuzzy Control Based Upon Man-In-The-Loop Identification." Doctoral Dissertation,1990, Rand Afrikaans University, Johannesburg, Republic of South Africa.
VON ALTROCK C:"Fuzzy Logic in Automotive Engineering." Embedded Systems Conference, 1994, 1995, 1996, Santa Clara, California, USA.

14 CONCLUSIONS

In this work, an attempt was made to equip the student as well as the practicing engineer with the fundamental principles and current design techniques related to fuzzy logic based industrial controllers. Those that wish to pursue more specialized research can avail themselves to the bibliography section which, however, by no means claims to be complete. Regarding classroom instruction, it was left to the discretion of the lecturer as to which sections to select for study for the particular educational level of his/her audience. In addition, the organizing of laboratory classes aimed at development practice and using advanced software tools, such as those mentioned in the text, would provide much-needed design experience and further insight into the design process and special design requirements of fuzzy logic based control systems.

A very important aspect of fuzzy logic design is to know in what applications fuzzy controllers can be used profitably. Experienced control engineers in companies regularly submitting proposals and bids for large-scale industrial projects should also include fuzzy control in their repertoire of design methodologies which can often place them ahead of their competition. At the same time, small enterprises could fill many technological gaps by manufacturing fuzzy controllers whenever large-scale mass production would not be justifiable.

Fuzzy and neurofuzzy control represent an emergent technology in industrial control, robotics, complex decisionmaking and planning. As such, it should also be included in the undergraduate curriculum of universities and higher schools of technology. This textbook, however imperfect, was written for this purpose.

INDEX